POLLUTED SITES
Remediation of Soils and Groundwater

POLLUTED SITES
Remediation of Soils and Groundwater

Paul Lecomte

A.A. BALKEMA/ROTTERDAM/BROOKFIELD/1999

Aidé par le ministère français chargé de la culture.
Published with the support of the French Ministry of Culture.

Translation of: ***Les sites pollués: traitement des sols & des eaux souterraines***, 2nd edition, 1998, © Technique & Documentation, Paris.

Translation editor : Dr. M.M. Oberai
Technical editor : Dr. G.D. Agarwal
General editor : Ms. Majithia

A.A. Balkema, P.O. Box 1675, 3000 BR Rotterdam, Netherlands
Fax: +31.10.4135947; E-mail: balkema@balkema.nl
Internet site: http://www.balkema.nl

Distributed in USA and Canada by
A.A. Balkema Publishers, Old Post Road, Brookfield, VT 05036-9704, USA
Fax: 802.276.3837; E-mail: Info@ashgate.com

ISBN 90 5410 784 7

Contents

I

Introduction to the Second Edition

Although the technical approach to study contaminated industrial sites, presented in this second edition, is identical with that presented in the first edition, it will be observed that the awareness at the level of authorities, mass-media, public opinion etc. has sharpened during the intervening years. The management and 'remediation' of polluted sites is today considered as important as the problems of air pollution, treatment of wastes, energy conservation etc. In November 1997, on the occasion of revision of the list of contaminated sites and 'dark points' (hazardous ones) thereof, Madame Voynet, Minister of Development of Land and Environment, France elaborated on the magnitude of the problem presented by polluted sites, declaring it to be her top priority concern. With this awareness, the concern for our heritage from the past becomes deeper: the list of polluted sites needing immediate attention, presented by the Minister, contains as many as 900 items. It may also be remembered that 20,000 hectares of brownfields contain many potential contaminated zones which have not yet been identified as such.

In the course of the last three years, a number of steps have been taken; some of them are listed below:

— setting up a committee for management of tax levied on specified industrial wastes for financing the rehabilitation of orphaned contaminated sites,

— elaboration by the Minister-in-charge of Environment of the policy regarding management of polluted sites, on the basis of risk assessment and publication of two detailed guides laying down simplified risk evaluation procedures (version 0, 1995; version 1, 1997),

— creation of a research institute dedicated specifically to polluted soils—The National Center of Research on Polluted Sites and Soils located at Douai comprising of about a dozen members from public and private sectors,

— etc.

Although the personnel specialized in financial and legal aspects, available for resolving the problem of polluted sites, have been significantly strengthened in France as well as abroad, there is need for further accelerating the development of technology for remediation of soil, particularly with scientific help from Northern Europe, the USA and Canada and enhancement of know-how on the part of French scientists and technicians. It is in view of this consideration that in this second edition Chapter 5, devoted to technology of remediation, has been expanded and rewritten with the collaboration of Isabelle Le Hécho and Fabienne Marseille. Conserving the pattern and style of the first edition, some techniques and new methods have been added. With respect to several 'standard' techniques, the developments and progress made during the intervening period have been included. Also, the chapter on costs has been revised and the changes in regulations included.

Introduction to the First Edition

Pollution...Environment...Ecology. These terms, relatively recent additions to our vocabulary (AFNOR, 1994), are today attracting much attention. Employed in every situation, even overused, sometimes hackneyed, they nonetheless represent more than media 'catchwords' or a passing fancy of public opinion. As a matter of fact, they make us aware of our society and, in a holistic sense, aware of the risks and hazards that technological developments during the latter part of the second millenium pose for the natural environment, our life style, and broadly speaking, our planet Earth. Considered today a highly diversified and environmentally rich inheritance, the planet constitutes a real 'gold mine' that is not to be squandered, but on the contrary, preserved and passed on to future generations. This awareness is quite recent, it has barely emerged; in fact even abuse of the environment is a relatively recent phenomenon since man's existence itself is but a dot compared to the long evolutionary period of our planet.

The 'biogeologic' system of the earth has been in existence for 3.5 billion years; it produced about 10^{20} tons of biomass (expressed as tons of dry matter) with no disequilibrium engendered in the natural milieu. On the contrary, its 'wastes' have been a source of wealth, for example deposits of combustible fossils (petrol, coal, lignite).

During the last 150 years the effects of industrialisation have ruptured the equilibrium which by recycling natural wastes had remained in balance over a long period of time. The sudden discharge of masses of toxic residues into the natural milieu has gradually led to numerous new risks to the equilibrium of the environment and ecosystem, as well as for man himself, the producer of these wastes and causative agent of this monumental disequilibrium.

Should further evidence be needed, let us recall that combustion of 5 litres of gasolene/petrol generates 20 grams of lead, which is sufficient to pollute 100,000 m^3 of air. The same 5 litres of gasolene spilled in a lake would spread over an area of 500 m^2 in the form of a thin covering layer, which would impede oxygenation and result thereby in the destruction of aquatic fauna and flora through gradual asphyxiation. One gram of hydrocarbons contaminates one cubic metre of water and our bones today contain, on average, 30 times more lead than those of Egyptian mummies.

This new awareness by society will certainly bring about a quick change in attitude towards the need for respecting and protecting the environment, leading to better health and well-being of fellow citizens. In all our activities and at every level of intervention (personal, collective and industrial) there will be an environmental concern in the design and execution of any new project.

Today, however, in keeping with this spirit, there is a need for considering the history and heritage of an exponential industrial development which has anarchic consequences for the environment and is often prejudicial for man and his natural inheritance. During the last few years attention has focussed on, among others, the risks related to contaminated industrial sites, either in use or abandoned, and a structured programme giving due consideration to regulatory and legal aspects has been set in motion.

This approach is of sequential nature (Fig. 1.1). The first step is to conduct a diagnosis, as accurate as possible, of the state and quality of the environment of the site under suspicion at a particular moment in its history. Subsequently, this state of the locale is evaluated in terms of the danger and risk for the community (persons working on the site as well as those living in the vicinity) and the natural milieu (surface and groundwater, soil and subsoil, flora and fauna, atmosphere, impact on climate etc.). Finally, if needed curative actions are proposed, directed towards rehabilitation of the site; they are properly planned and executed after assessing their technical and financial feasibility. The implementation of such remediation is reviewed and procedures revised, if so warranted, at various stages of the investigation and curative action. Being interdisciplinary in nature, this process requires a rigorous and structured approach; this is what we have endeavored to achieve in this book.

Our objective in this book is a succinct presentation of the different phases of the sequential approach mentioned above. We shall focus on 'remediation' and consider specifically two natural media of the environment: soil and groundwater.

Chapter 2 reviews contaminated sites and discusses their classification based on different criteria: type of pollutant, activity at the site etc. The next two chapters are devoted respectively to pollution diagnosis (Chap. 3) and assessment of the risks involved (Chap. 4). These two steps in the chain of investigation and treatment are indispensable to an understanding of the subject of remediation per se. The process of remediation is detailed in Chapter 5; attention is paid in particular to the criteria of choice of technology and feasibility of implementation on the one hand, and classification of techniques into 'families' on the other. Another chapter is devoted to estimation of the costs of implementation of the remediation measures (Chap. 6). Finally, the last two chapters synthesise the legal aspects (Chap. 7)

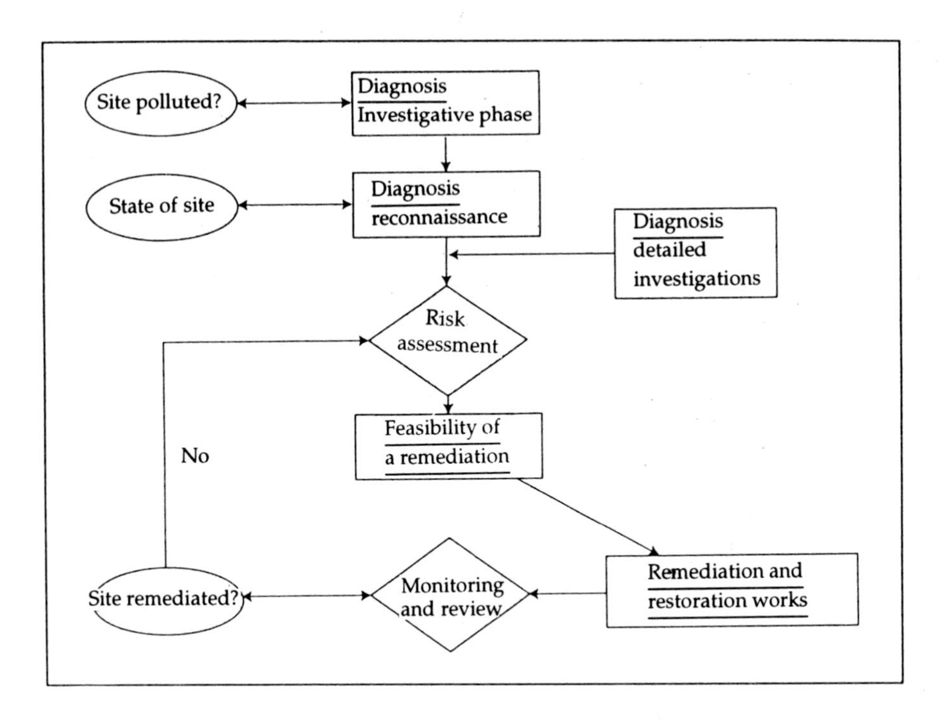

Fig. 1.1: Scheme of the sequential approach to investigation and treatment of a polluted site.

and considerations of insurance (Chap. 8) specifically as prevailing in France.

The book is primarily addressed to those concerned about the quality of life, who desire to understand and remedy the prejudicial consequences of industrial activity on the environment. It is also addressed to the personnel in charge of a site, an activity or an enterprise, as well as to all technicians and engineers responsible for the environment in the course of their work. Lastly, it is addressed to all those who have framed environmental regulations or have suggested means for their enforcement.

Before embarking on the subject proper, I wish to thank my colleagues of the Directorate of Environment of ANTEA for their judicious advice and support. I am grateful to Jacques Ricour who reviewed the entire manuscript and collaborated specifically in the final version of Chapters 2 and 8. My sincerest thanks are due to Liliane Laville-Timsit and Bernard Côme for reviewing Chapters 3 and 4. Finally, Claudine Kluyvers and Jean Pierre Samour handled the responsibility of graphics; I gratefully acknowledge their contribution.

II

Contaminated Sites

2.1 MAGNITUDE OF THE PROBLEM

Only very recently has France become aware of the importance of contaminated sites, particularly their number, the area covered and their diversity. In comparison the USA, the Netherlands and the (former) Federal Republic of Germany conducted a census more than ten years ago of sites likely to pose environmental problems.

The last two decades have seen a profound change in attitude in France, which has affected her social and economic fabric. Deep concern over the impact of industrial and service activities to the detriment of agriculture has developed. Migration of the population to towns and cities has accelerated desertification of the countryside.

One outcome of this change in attitude is the reconversion of industrial sites for reasons of safety, dilapidation and human occupancy. Such reconversion has brought about awareness of the inevitable difficulties encountered, for example, in the case of sites at Montchanin, Sermaize, the mine-fields of Lorraine, St. Etienne and Nord-Pas-de-Calais, the textile-rich areas of Cambrésis or Vosges and others.

Thus, under pressure of public demand inventories of 'hazardous sites' were prepared in the years 1975–1980. To assess the magnitude of the problem, ADEME (see Table 7.2) has been preparing a systematic inventory of the French regions since 1992, based on archival records. In 1993, the first phase of investigation was completed for the Toulouse region; such investigations are underway for numerous other regions, for example Lorraine and the Great Parisian Belt.

It presently appears that the number of contaminated sites is far larger than had been imagined. Among the industrial wastelands (20,000 hectares according to the 1987 census), refuse dumps for domestic wastes (6300 as per the census) and hazardous sites of dubious or unidentified origin, there are a few which pose no difficulty from the point of view of rehabilitation, taking into account new usage of the site and level of degradation of the soil and groundwater.

According to the recent (1993) estimates of the Bureau of Frost and Sullivan, the number of contaminated sites in France could be of the order of 100,000, of which 20,000 are slated for rehabilitation.

There is also an urgent need to undertake an elaborate census for the purpose of typifying the sites on the one hand, and ranking the actions to be taken on the other.

2.2 WHAT IS A CONTAMINATED SITE?

In connection with elaboration by census of a typification and a method of ranking contaminated sites, the question arises: 'What is a contaminated site?'

A site which is geologically known to contain metallic sulphides, a granite which naturally emits radiation or a marl loaded with evaporites are excluded from the definition since local concentrations of certain chemical elements (minerals, organic compounds or nuclear radiation) could have been generated by natural processes.

A contaminated site is thus one that assumes a human activity which has resulted in the degradation of the natural environment (soil, subsoil, groundwater) to the extent of rendering it unfit to a new degree and/or transforming the order of risks for the inhabitants or the users.

A refuse dump from which mercaptans, methane or solvents escape would thus be a contaminated site in the same sense as an old mine-field whose soil has been polluted with PAH (polycyclic aromatic hydrocarbons), phenols, or metals; or an abandoned tannery or a factory, in or out of use, if it discharges effluents (liquid or gaseous) or untreated wastes into the natural environment.

It follows from this viewpoint that a site used for a specified period of time corresponds to a degradation of the surroundings: the countryside, biological diversity, physical or chemical quality of soil and water, and so forth. Prevention, right from the stage of designing the activity and preventive maintenance, is thus the only means of avoiding or limiting deterioration of our environment. Furthermore, economic constraints arising from the prohibitive cost of remediation should not be allowed to stand in the way of dismantling installations at the end of their lifetime and putting the space made available thereby to new use.

A contaminated site may be defined as a space wherein the activity of production, transformation, transport or service etc. is carried out and which, due to negligence or defective design or improper maintenance, leads to the occurrence of damage and immediate or deferred risks for the users, the present and future inhabitants and for the environment (Ricour, 1993).

It is important to emphasise that the above definition includes the immediate potential dangers as well as those deferred in time or space by the existence of natural or artificial factors of 'retardation'.

2.3 CLASSIFICATION OF CONTAMINATED SITES

Elaboration of a typification of contaminated sites (inventory, classification and ranking) has thus become necessary for itemising the cases that occur and for a better understanding of the origin and the phenomena encountered.

The variedness of the cases encountered necessitates an ordering directed towards precising the specific methodologies applicable to each large class thus defined, in order to reduce the cost of intervention and to employ methods which are as objective and identical as possible, thereby precluding the introduction of a competitive basis from one region to another or from one country to another.

The absence of such typification and methods of approach would, as a matter of fact, lead to large distortions in considering the cases of pollution in terms of the culture of each region, constraints of lay-out, availability of land etc.

Contaminated sites can be classified according to various criteria: legal standards, nature of pollutant(s), mode of dispersion of contaminant, type of activity, number of pollutants and so forth.

2.3.1 Classification According to Legal Standards

— European and recent (1992) French laws adhere to the policy 'polluter must pay'. In the case of certain sites which were active long ago and have long since been abandoned, it is not always possible to fix legal responsibility for the contamination; the proprietor of the land can neither be held responsible for 'past events' nor may be solvent in many cases, given the difficulties involved.

Contaminated sites of this type are called '**orphaned hazard sites**'. Taking charge of them and restoring them comes under community responsibility with appropriate technical, legal and financial duties.

— The second type of contaminated sites comprises **industrial wastelands**. It encompasses such industrial terrains or zones as were abandoned following cessation, usually sudden, of an activity or closure of a regional, national (even international) branch of industry. Slowing down and stoppage of these activities are related to many causes: very strong pressure of international competition, worn-out tools of production, technological progress which has rendered certain procedures obsolete, geographic enclavement, very high cost of labour, energy or primary material and so forth.

Industrial (as well as harbour, railway, military etc.) wastelands are large. The suddenness of their appearance and the magnitude of the problems posed for environmental as well as sociocultural and economic rehabilitation often necessitate a cooperation of all parties concerned: industrialists responsible for the activity, the local community, the state in the case of public or semi-public undertakings and above all entities like the European Union in the framework of specific plans such as the FEDER or CECA.

The collective charge of environmental consequences for the foregoing is not as well developed as in the case of orphaned hazard sties. As a matter of fact, if the industrialist is to be regarded as the principal actor, the problems to be resolved necessitate, as in the case of orphaned hazard sites, a large association of all parties concerned with the choice of rehabilitation in the framework of an 'open' communication.

Finally, contaminated sites may correspond to an **ongoing industrial activity** in which the proprietor is unambiguously identified as the person responsible. The new regulations require, as a first step, installation of devices for prevention of harm to any third party and to the environment exterior to the site. As a second step, there must be provision for resorption of harmful products and for remediation of the site.

The proprietor holds the reins of communication with the other parties concerned: administration, inhabitants, bankers, insurance companies, ultimate buyers of the site; he is also responsible for compiling information that must be made public as per the existing laws.

2.3.2 *Classification According to Dispersion of Pollutants*

Another type of classification distinguishes contaminated sites according to the manner of dispersion of pollutants in the natural surroundings (Fig. 2.1).

The figure presents two distinct cases:

— Diffuse contamination resulting from the site activity itself. The pollutant steadily spreads in small quantities into the surroundings, sometimes over a long period of time (many years). This is the case, for example, of percolation and leaching of pollutants to underground channels or flow of pollutants into the surface-water network, or fallout of dust or smoke. Most often this is **chronic pollution** which is detectable only after considerable time has elapsed.

— Sudden en masse pollution, corresponding to the discharge of a large quantity of pollutants in a short period of time. This is termed **'accidental pollution'**. Thus an overturned tanker truck or explosion of a tank or a reactor belong to this category.

In both cases the form of the source of pollution constitutes an element of fundamental classification:

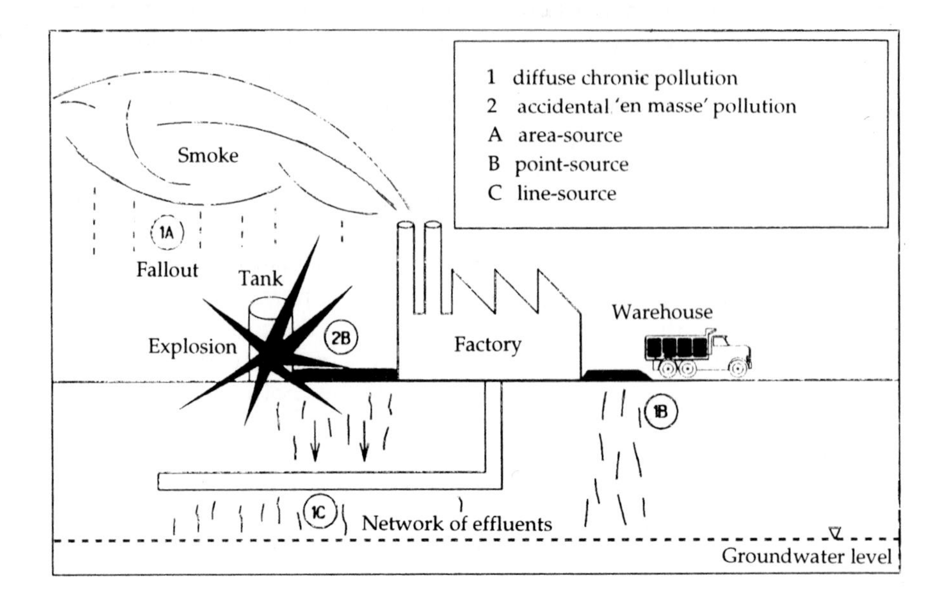

Fig. 2.1: Diagram of types of contamination based on mode of dispersion of pollutants.

— The source of pollution is an 'area-source' when spread over a large area or the contaminant introduced into the surroundings is distributed over a large space. A typical example is related to the fallout of smoke (with a concentration of contamination which decreases as one moves away from the point of emission).

— The source of pollution is a 'point-source' when the contaminant is introduced at an exact point and restrained in space. This is the case, for example, of a leakage at the level of a storage tank.

— The source is a 'line-source', horizontal (waterways, network of underground or surface channels) or vertical (empty well, drill hole, mine shaft) when the contaminant is distributed along the route while transporting the pollutant from one place to another by a device with a chronic defect in watertightness or even accidental rupture at many points.

2.3.3 Classification According to Situation Detected at the Source or Impact on the Targets

Though this classification is relatively simple, it is not less important since it implies progressive investigation and suggests a plan for implementation of various curative measures.

This classification is based on the separation of pollutants detected directly **at the source**—group of industrial sites—and those detected **on the**

target where contamination poses a direct danger for the population and the ecosystem—classic case of contamination of potable water collection system or a waterway.

In the first case one defines the cause and in the second, the effect. In the first case one knows the source of contamination and very often also the person responsible for it. In the second case, initially one sees the damage caused to the environment, most often only in a crisis situation (as in the case of a drinking water collection system) and knows neither the source nor the person responsible. The approach would thus consist of imposing urgent safety measures and concomitantly making an effort to discover the source of contamination by gradually moving upstream.

2.3.4 Classification According to Nature and Number of Pollutants

The nature and number of pollutants are complementary elements which allow the ranking of contaminated sites in terms of the types of activities in which they are engaged: steel industry, petrochemicals, light chemicals, heavy chemicals, metalworks etc., or in terms of types of particular sites: marshalling yards, military activities, storage of transit goods etc.

In France one adopts the APE classification of enterprises elaborately described by INSEE, which offers a homogeneous division of the various existing hierarchies.

This classification can be further extended by distinguishing between:

— a **single-** or **multiproduct** contamination. Thus, contamination would be termed 'single-product' when it pertains to contamination by a single contaminant, e.g. in the case of the subsoil around a petrol or gas-oil depot, and 'multiproduct' when it pertains to the site of a chemical industry producing a range of products which are potential contaminants;

— and **organic, mineral, radioactive** or **mixed contamination**. A typical example of mixed contamination is the site of an old gas factory in which heavy metals and cyanide may be associated with phenols and polycyclic aromatic hydrocarbons (PAH).

III

Environmental Diagnosis

3.1 DEFINITION AND OBJECTIVES

3.1.1 Definition

Environmental diagnosis constitutes the first step in studying the contamination-remediation of a given site. It represents the phase of initial investigation necessary for acquiring knowledge as to the state of the environment of a site. There will always be interest in the investigation being as detailed as possible, with the maximum information gathered before embarking on the subsequent steps, namely evaluation of dangers and implementation of operations of remediation (see Fig. 1.1).

The essential purpose of an investigation of this type is to characterise the environmental state of the site at a moment of its history or, more concretely, to quantify the surrounding media, natural or otherwise. Mostly, the investigation would concern the soil, subsoil, groundwater or even air and sometimes also the surface water (for example, a river close-by), the fauna and the flora, the state of health of the inhabitants and so forth.

The success of a diagnostic mission depends on the fulfilment of certain conditions, generally stated in a written document with precise definition of:

— the parameters of intervention in time and space by the person in charge of the study;

— the chain of communication to be obeyed in terms of level of confidentiality vis-à-vis various participants;

— the documents and sources of information available in written or oral form.

In many cases the investigation will be conducted in an indirect manner, i.e., not by directly characterising the milieu, but by searching for the eventual contaminating products which could penetrate into, accumulate at or migrate towards the boundary of the site. Thus, for example (Fig. 3.1), during a study of the environment around a factory manufacturing pesticides, one would quantify at the level of different media (atmosphere,

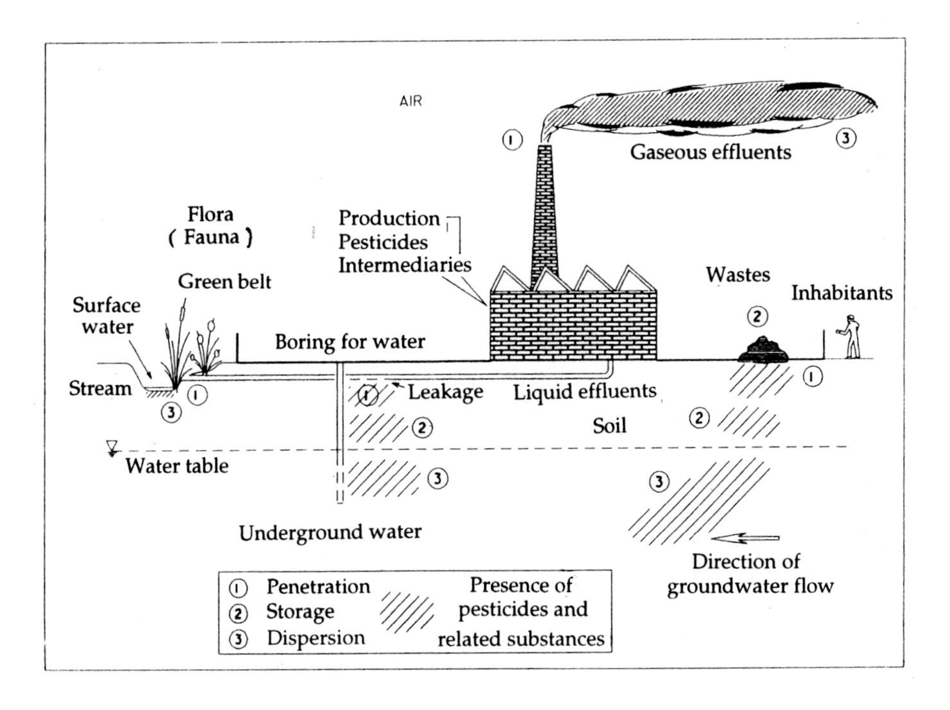

Fig. 3.1: Investigation of a polluted site: the relevant media.

soil, groundwater table etc.), the eventual residues of the primary material used, the pesticides manufactured, the by-products or wastes generated during manufacture, rather than try to characterise the global composition, mineral or organic, of each medium under study.

In this perspective there are three major steps of investigation characterising a diagnosis of the environment, namely:

1. **Identify the ultimate pollutants and characterise them** in terms of concentration, volume, form (pure products, salts etc.) or phases (gases entrapped in soil, soluble in groundwater etc.):

— localise their occurrence (around such and such workshop, under such and such tank etc.);

— define their gradient of distribution (dispersed on a surface of such and such size in such and such quantum).

2. **Determine the source and cause of pollution:** is it an accidental or chronic pollution? In either case can the duration of the phenomenon be estimated? Is it a concentrated or diffuse pollution? To which process is the phenomenon related: temporary storage of primary material, discharge of effluents, inadequate treatment of gaseous emissions etc.? How has the pollutant reached the medium and to what extent has it spread?

At this stage of the study there are several questions and several possible answers: they sometimes require a circumstantial inquiry into knowledge of the site and its activities or, preferably, an actual dialogue with the present and past exploiters of the site.

Quite often, taking note of this point at the first instance expedites resolution of the question of the nature of the pollutants.

3. **Characterise the physicochemical and hydrogeological conditions of the site** which will facilitate determination of the vulnerability of the various media. What is the composition of the subsoil? Is it a medium permeable to seepages (a liquid pollutant spreads more readily in a fractured medium or sand than in clayey sediments)? Is it protected on the exposed surface (by a lining, a layer of asphalt etc.)? Does an aquifer exist in the area? At what depth is it located? What are its characteristics? For example, what is the direction and velocity of groundwater flow in the aquifer? What are its uses? and so forth.

3.1.2 Objectives

Diagnosis of a contaminated site will thus have to fulfil the following objectives:

• Provide a contractual statement, as accurate as possible, of the degradation of the natural media—water, air, soil and subsoil—which will be acceptable to all parties (manager of the site, administration, buyer or seller, insurers, bankers, communities, associations etc.) depending on the external references to the significant and recognised site.

Thus concentrations of lead (Pb) or zinc (Zn) of the order of 1000 mg/kg observed in soil around a factory engaged in treating ores would be considered abnormal compared to the quantum of 100 mg/kg found naturally in similar soils of the region. Such concentrations would constitute a pollution derived, for example, from the fallout from the smoke of the factory (Fig. 3.2). On the other hand, a quantum of 150 mg/kg arsenic (As) would be perfectly normal in a region where the geological layers are naturally rich in metals and are used for mining.

• Constitute a necessary and sufficient information base which could serve as a reference for:

— Aiding the **choice of a future method of remediation** by taking into account all other factors of decision (see Chaps. 4 and 5.1). In the case of pollution with Pb-Zn cited above, the diagnosis would enable, for example, delimitation of the zone in which the soil would have to be excavated to extract the heavy metals.

— Judging the **efficacy of these means of rehabilitation** implemented in accordance with the required objective of remediation. After extraction of metals from the contaminated earth a new analysis would reveal the residual concentration (for example 150 mg/kg); it would thus be possible

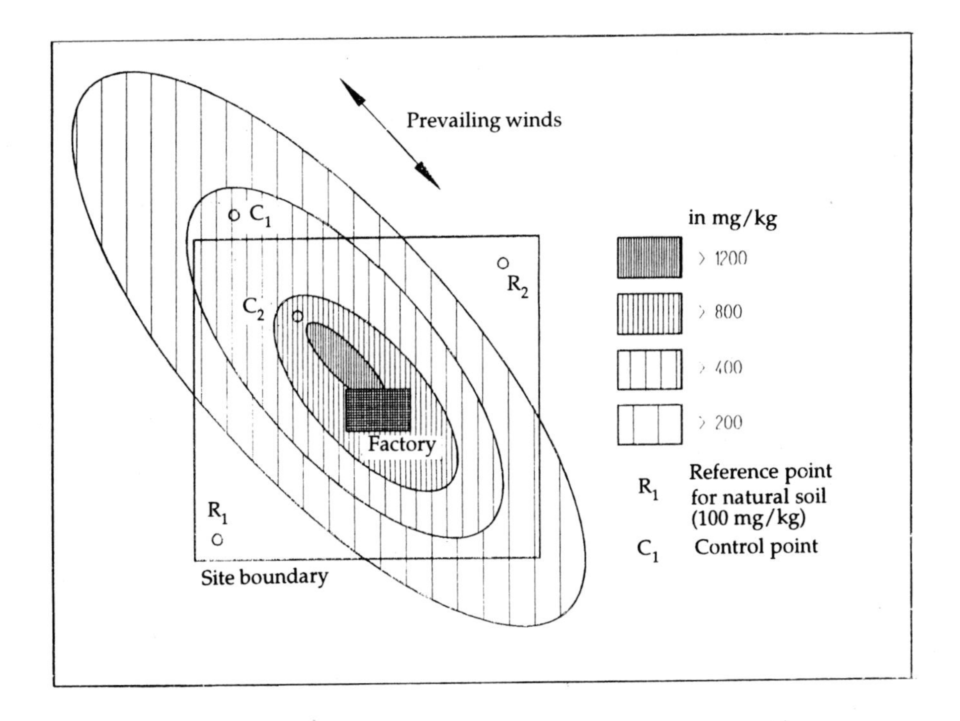

Fig. 3.2: Example of distribution of lead pollution around an active site.

to estimate the output of the remediation vis-à-vis the quantum observed initially (1000 mg/kg) and to ascertain whether the operation has attained the goal set in terms of concentration among the concerned parties (in this case 200 mg/kg).

— Elaborating the conditions of hygiene and safety during the work as well as **a system of follow-up and control of the site** in accordance with the laws in force. In our example the industrialist would have to install a system of emission treatment at the outlet of the stack to decrease the quantum of metals present in the smoke. A check on the concentration of metals in the soil around the factory would enable verification of the efficacy of the system over the course of time compared to the quantum observed initially (Fig. 3.3).

3.2 NATURE OF OBSERVATION

Whether or not its execution by the proprietor or an employee is voluntary, a diagnosis of pollution does not constitute a value judgement on behalf of the person in charge of the study; it is rather a 'picture' at a precise moment of the state of the natural surroundings of a site.

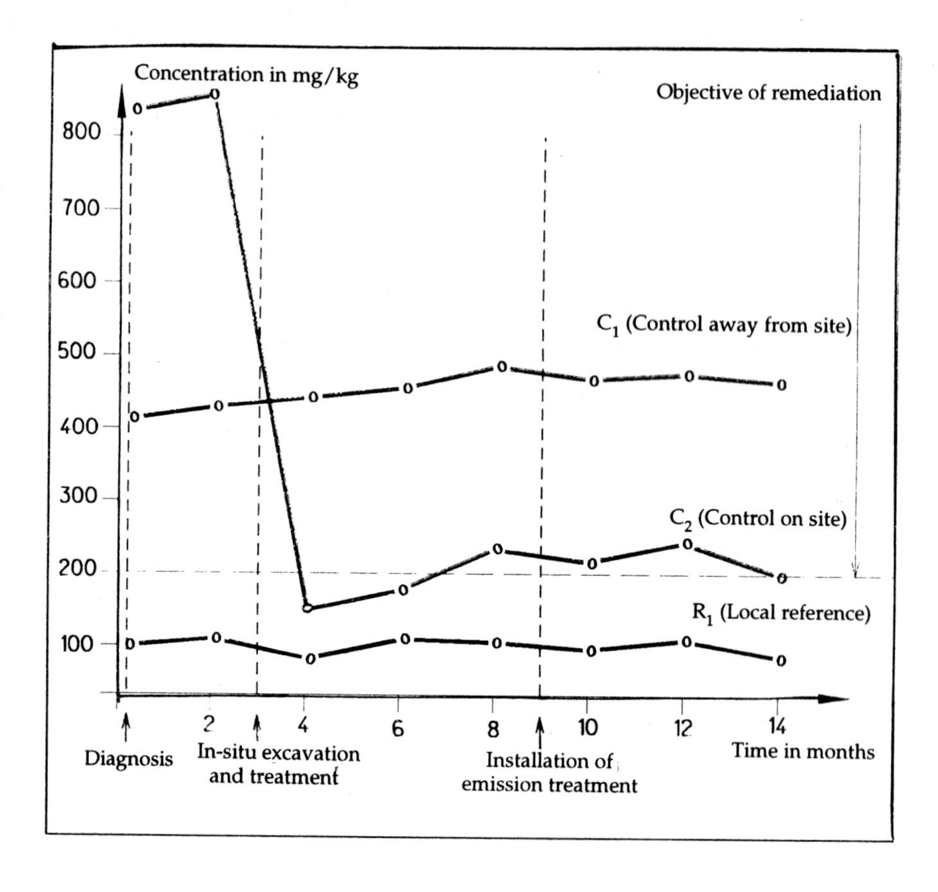

Fig. 3.3: Development over time of concentrations of lead in the soil at the control points.

Thus the person in charge of the investigation **has to be factual in his observation**, validating and checking his sources of information and the data obtained, and detailing in the form of a final document:

— procedures of measurement and intervention,
— validated factual data,
— results of computation,
— interpretation.

Thus, rather than using such expressions as 'highly polluted sample', 'strong pollution' or 'extremely contaminated zone', which in themselves are meaningless, it is preferable to state facts by comparing the quanta observed with one or several initial reference values, for example 'quantum of lead in the soil of the site is 1000 mg/kg with a background level of 100 mg/kg in the environs'.

Furthermore, the person in charge of the study has an **obligation to advise and maintain contact** with his interlocutor, especially in the matter of the existing regulatory restrictions, by supplying him with the material needed for a clear appreciation of the results obtained, and proposing technical and administrative solutions which are not only appropriate, but also economically feasible.

Another ethical aspect to be taken into account in a diagnostic study is the **obligation of confidentiality** with regard to the results of the study as well as the manner in which it is executed. As a matter of fact, aside from the duty of maintaining secrecy about all studies undertaken for a client, the diagnosis of environment sometimes touches upon a difficult social framework, for example during redemption of the enterprise. More generally, the site or the activities of an industrialist may be subject to pressures from ecological movements, often radically reported in the communication media, sometimes to elicit a predetermined impact on public opinion. Such contexts necessitate a total confidentiality during investigations, which is sometimes difficult to maintain, especially when legal implications are involved.

3.3 WHY A POLLUTION DIAGNOSIS?

A request for diagnosis of pollution is seldom made without an a priori or initial suggestion or apprehension; it implies a certain stand taken by the exploiter of the site—or the proprietor—or perhaps the administration. It also implies technical and financial aspects and very often understandably some communication and exchange of information among the various parties concerned.

So, at the outset one should try to ascertain the reasons which prompted the proprietor of a site to requisition a diagnosis. In what cases would the undertaking of such an investigation be warranted?

Although the idea behind a diagnostic observation is always the same, the motives of the owner or the exploiter of the site may be varied:

— Study of Impact

Carried out in a strict regulatory framework, the aim of this type of study is to determine the consequences of a new activity—or change of activity—for the environment and the inhabitants. It is an environmental investigation, always conducted as a basic requirement and not because pollution is suspected. It is more or less a 'zero-state' of environment at the initiation of an activity or a modification or extension of a classified establishment (see Chap. 7).

— Monitoring of pollution (episode, complaint ...)

This may be the more obvious case: after report of a pollution episode or on a complaint by a third party, for example of some nuisance caused to the inhabitants near an industrial site, monitoring of pollution is undertaken to determine the origin, causes and effects of the episode or the stated nuisance.

— Transfer/Sale of an Industrial Site

Not widely practised in France until a few years ago, a diagnosis of the state of environment is now carried out for every site which is to be sold or given away to a third party. Termed a 'technical audit of the environment' in this particular case, this type of investigation has also been a requirement since 1992 as per legislation under a law pertaining to contaminated sites. It obliges the seller of an industrial site to make the buyer aware of the actual state of the environment of such a site at the time of transaction (see Chapter 7, Legal Standards).

— Conformance to Regulations

In the domain of environment, laws evolve rapidly and industrialists are today confronted with frequent changes and evolution of the legislation in force. Furthermore, depending on the modification of site activities, the administration can order an industrialist to review and improve the plan of protection of his immediate surroundings. In certain cases the industrialist has perforce to arrange an investigation of his site, or a part thereof, usually with a precise objective of ensuring compliance with new legislative requirements. He has, for example, to drill a borehole downstream of the zone under consideration to provide a check on quality of the groundwater in reference to certain chemical parameters.

The above cases represent most of the requests for intervention but other types of situations necessitating an environmental diagnosis may also occur. Thus,

— obtaining certain bank credits,

— establishment of reference for assessing insurance premiums,

— laying down policy for protection and prevention of environmental degradation at the level of an industrial group or an ensemble of sites,

— installation of environmental indicators for a site, which can be utilised in the scheme of communication within or outside the enterprise— may necessitate investigations of the state of environment.

3.4 CONDUCTING AN INVESTIGATION

Conducting an environmental investigation and a diagnosis of contamination of a site calls for concepts which are sometimes complex in terms of basic principles but which have resulted in simple schemes currently in use (Barrès-Lallemand, 1993; Pellet and Laville-Timsit, 1993).

Actually, the large diversity of potentially contaminating products,[1] their eventual mutual interactions and their dispersion in widely varied

---

[1]More than 100,000 organic products are commercially available throughout the world today; in addition, 3000 to 4000 new substances appear on the market every year, excluding the products of degradation.

media (varied geological and hydrological situations) do not permit planning an approach adapted to each individual case. On the other hand, institutions engaged in specialised design have shown a tendency to develop a more general investigative approach, which is sequential in nature and adaptable to all cases.

This approach is multiphased, starting with simple and general steps and moving towards more detailed and complex ones in terms of methodology as well as the information expected. For a given site, the approach is centripetal, initially envisaging the site in its totality to obtain overall knowledge, then focussing through a series of steps more and more precisely on locations concerned with particular activities and products.

There are usually two distinct phases of investigation:

— a survey of 'background' consisting of gathering all the known information regarding the site, including its history (phase 1); the term '**historical survey**' may also be used but such usage is rather restricted (see below);

— a study termed '**study of the terrain**', which may include analysis of certain samples from different media conducted in situ or in the laboratory (phase 2).

The latter phase can be subdivided into phase '2a', relatively global, taking into account the ensemble of the site and the zones potentially very suspect, and phase '2b' (and eventually '2c', '2d' ...) related more precisely to certain parts of the site to be studied in detail for different materials depending on the results obtained during the preceding steps.

After synthesising and interpreting all the data gathered, an expert report is prepared, which presents the pollution diagnosis as well as an analysis therefrom of the hazards and risks inherent in the site vis-à-vis the neighbouring population and the local ecosystem (see Chap. 4).

In practice, the requisitioner of such an investigation sometimes evinces interest in an analysis of preliminary hazards upon completion of the 'background survey phase' but prior to the 'terrain study phase' of the diagnosis. Actually, depending on the information obtained during the background survey (phase 1), the investigator could come to know that in certain cases damage to the environment and hazard for the inhabitants are negligible, in which case phase 2 need not be undertaken.

3.5 MAJOR METHODOLOGIES

3.5.1 Phase 1: Survey of Documents

It is commonly accepted today that during an investigation the goal of the survey phase is to collect and synthesise all the facts known about a given site, drawing on all sources of information—written, graphic or oral. The scope of this document survey and the sources of information to be tapped are defined beforehand in time as well as space and require no technical work of sampling or chemical analysis. The synthesised data is validated and generally represented in a homogeneous and coherent geographical format (data plus cartography). This report concludes with proposals and concrete recommendations either for an analysis of hazards or a plan of investigation with sampling and chemical analysis.

All participants—investigators, administration, proprietor, works manager etc.—admit that this document-survey phase is indispensable today for:

— limiting the field of research in later stages of investigation;

— defining the optimal conditions of safety for future works.

Sometimes called a 'historical analysis', phase 1 of a diagnosis is indeed more extensive than a historical survey of past and present activities of the site. As a matter of fact, one can integrate into the background-survey phase all the general information of the site and its natural environment, which constitute the second phase.

• **Historical analysis**

Historical analysis considers all activities on the site, both present and past, irrespective of whether they involve the present industrialist or past exploiters. It must take into account all processes and operations followed in the course of activities at the site from its debut to the present day. For a site with a long industrial tradition, retrieving all the documents concerning the earlier activities is no easy task; some sites are more than a hundred years old and records relating to the period before the Second World War are scarce and eyewitnesses of site activities rare.

Research and compilation of material balance-sheets. These are always envisaged in relation to the materials used. The purpose is to determine the types and quantities of materials involved in a given process. Thus one tries to ascertain the primary materials used, the intermediates formed and the final finished products. At each step one strives to determine whether the system generates (generated) wastes and the amount of such wastes, either in the form of gaseous or liquid effluents or in the form of solid wastes. In most cases it is possible to make a qualitative and quantitative balance- sheet of input and output of materials.

This part of the survey enables preparation of the inventory of known products, potentially contaminating substances which have transited on the site.

The second aspect of the survey of the activities is related to the **disposal of wastes** generated during the manufacturing process. Thus it is essential to determine the customs and procedures regarding the removal of wastes of all forms and to ascertain the transport channels currently used. The presence of an old internal dump, network of effluent drains etc., may necessitate a long and patient investigation. A detailed reconnaisance at this level enables identifying potential dangers existing on the site vis-à-vis the environment.

Broadly speaking, the historical survey represents a difficult task whose profitability depends on an **actual dialogue with the exploiter** and his colleagues. In some cases this is none too easy; the exploiter may not be the direct requisitioner of the study and thus may have little interest in the survey undertaken. On the other hand, even if he has requisitioned the study, the proprietor of a site generally tends—all in good faith—to under-evaluate the impact of his activities on the environment and its eventual harmful consequences. This is particularly true for old activities when it was not customary to take precautions vis-à-vis the environment. The common practice, for example, was to let the liquid effluents discharge directly into the subsoil and allow the solid wastes to accumulate on the immediate boundary of the factory, with no safety measures taken whatsoever.

Sources of information to be considered are multiple. First and foremost are the records kept in the factory (current dossiers or archives). Different types of plans, maps, photographic material (old commercial plaquettes, aerial photos taken at different times etc.) are generally of great importance for reconstructing the history of the site. All official or administrative documents (police decrees, dockets of removal of wastes etc.) are equally important for yielding information as to the various environmental criteria earlier applicable to the industry.

To illustrate the multiplicity of possible sources of information, Table 3.1, though not exhaustive, lists the documents which, if available, should be collected and consulted.

One should also include in the survey a questionnaire for the personnel associated with the site, e.g. those directly concerned with the protection of the environment, the persons in charge of the effluent treatment plants, (laboratory analysts etc.), those in charge of production processes or some specific activities (logistics, management of space) ... etc. Discussion must be carried out with the permission of and in consultation with the exploiter or his representative, who has a major stake in satisfactory progress of investigations. It is also often useful to appeal to old members of the staff, in service or retired, to clarify a current or old situation from their memory and experience of the past.

- **General information about the site**

Phase 1 of the site diagnosis obligatorily includes a detailed visit to the site. This actualises the situation only partially discovered from the documents. Particular conditions will be encountered whereby one learns in detail the potential impact on the natural surroundings, observes the general cleanliness, detects any blights on the soil, and can examine piling of wastes, and the state of pools and effluent channels etc.

Table 3.1: Examples of documents to be taken into account during the documentary-survey phase of diagnosis of a site

1. DOCUMENTS RELATING TO REGULATIONS
   - Police decrees or administrative directions relating to the site (including classification of installations for treatment of wastes on site, remediation plants, incinerators etc.).
   - Land-use and zoning.
   - Risk assessment and management plans.

2. CARTOGRAPHIC DOCUMENTS
   - Location of the site in relation to its immediate surroundings.
   - Plans showing
     - locations of buildings
     - underground and surface pipe networks
     - old and present storage areas
     - location of surface and underground reservoirs
     - location of any embankments, earthfills, borrow pits, etc.
     - location of wells, boreholes etc.
   - Site plan showing asphalted areas, uncovered soils, turfed patches, areas covered with gravel etc.
   - Plans and operating schedules of installations for treatment of solid wastes (incinerators, compactors etc.) and of effluents treatment plants.

3. MISCELLANEOUS
   - Photographs (aerial if any) retracing the history of the site.
   - Description of various sections of terrain, boreholes, wells etc. that can provide information about the nature of the soil and subsoil of the site.
   - Documents, plaquettes, newspaper articles etc. retracing the history of the site and its evolution over time or relative to its usage before being taken over for development.
   - List of primary materials and other substances used on the site for manufacturing finished and intermediate products.

A visit to the site also provides information that is useful in preparing the plan of the 'terrain study' including sample collection, depending on the zones of potential risks and the safety norms to be followed.

During the visit, supplementary questions can be put to the exploiter or his representative, yielding more detailed information. Photography, with prior permission of the industrialist, is an excellent means of illustrating various points when the final synthesis and recommendations are presented.

General information about the site also involves researching documentation on the natural media. This would provide detailed information relating to soil structure, description of earthwork and drilling etc. carried out at different periods for various purposes like foundation studies, search for water, demolition or construction of new works (see Table 3.1). Technical reports on the state of the subsoil and its geological or hydrological characteristics are of primary importance. Information of this type can be gathered from the National Geological Service (regional office, databank of subsoil etc.) which has conducted local and regional studies.

This part of the survey enables characterisation of the subsoil and assessment of its vulnerability vis-à-vis eventual aggression due to activities at the site. Supplementing this with knowledge of the site surface (presence of concrete surfaces, system of water drainage etc. as well as the influence of climate, wind etc.), it becomes possible to determine the most vulnerable zones and routes of pollutant-transport into the natural media in advance of this actually happening.

Figure 3.4 shows the principal elements to be taken into account for determining the preferential circuits for a natural milieu that is more or less vulnerable:

— lay-out and surface characteristics of site: concrete-lined or turfed surfaces, gravel embankments, drainage network for storm run-off and effluents...;

— types of soil and succession of geological strata with their characteristics: physical and chemical composition, permeability, compactness, thickness ...;

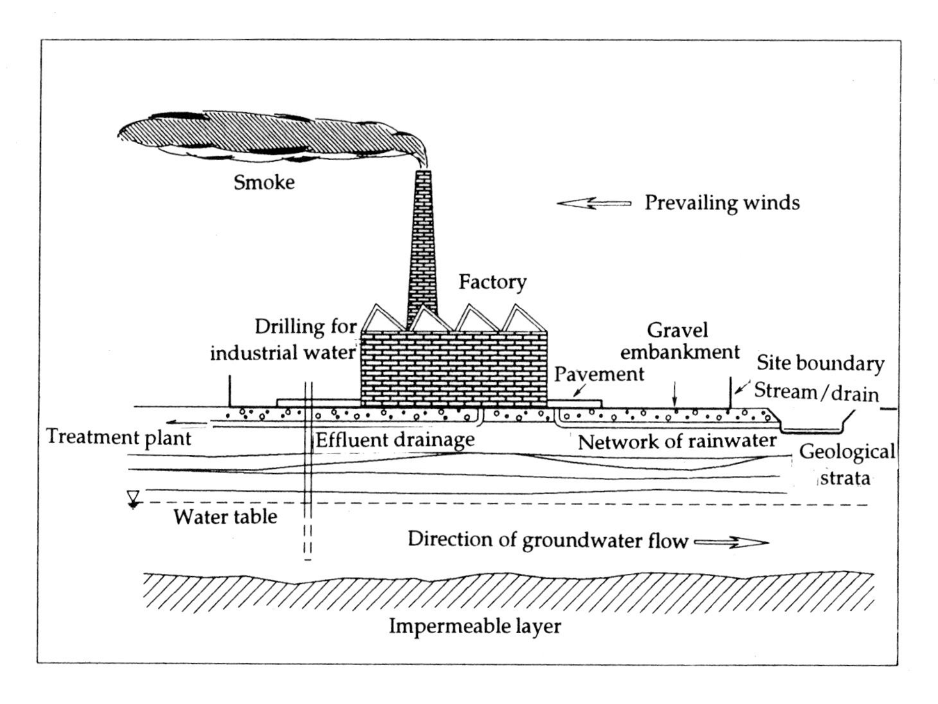

Fig. 3.4: Factors influencing the pollution-transport routes and vulnerability of the natural milieu.

— presence of a water table, depth, gradient and velocity of flow, eventual usage for potable and industrial water;

— climate: rainfall, prevailing winds, fog etc. and its impact on the meteoric conditions of the soil.

3.5.2 Phase 2: Terrain Study

After completing the first phase of the diagnosis, a programme of detailed investigation of the site is implemented. This programme, prepared in the course of phase 1, is designed by considering:

— information gathered during the phase of survey of documents;

— objectives specified by the client;

— directives and guidelines communicated by the administration or by legislation regarding the controls to be provided on the site, the permissible limits of contamination etc.

The objective of the second phase is to confirm the presence of suspected contamination by quantifying the degradation observed in the natural surroundings in terms of type and concentration of contaminants, and significance and extent of their dispersion. If this investigation is sufficiently detailed, it becomes possible to estimate the extent of contaminated terrain and the distance over which pollutants have spread—the data necessary for subsequent operations, such as analysis of hazards and choice of methods of remediation (see Chaps. 4 and 5).

A number of methods are available for attaining this objective; they are classified into two groups:

— direct techniques of in-situ monitoring (geophysical techniques, hydrogeology, physicochemical monitoring of the soil, water or air, infrared photography ...);

— methods of indirect investigation involving sampling and laboratory analyses, often away from the site (geotechnical, chemical ...).

By and large each site represents an individual case in which eventually various methods may need to be combined with appropriate weightage depending on the characteristics of the site and specific goals of the investigation. Thus an accidental pollution of the soil by petrol could be characterised by directly monitoring in situ the concentration of hydrocarbon gases present in the soil. The presence of waste-casks buried in an abandoned industrial site can be spotted on site by a battery of georadars. Or again, the spread of trichloroethylene contamination into groundwater can be detected by drilling downstream where a water sample is taken and sent to the laboratory for analysis of the chlorine compounds.

The optimal methods of monitoring to be employed depend on a number of factors which are difficult to generalise and always mandate strict rigour in the approach.

Further, depending on the required level of investigation, the importance of the techniques to be employed will vary. A high degree of precision necessarily involves more sophisticated—and expensive methods.

For a less intensive investigation, a first estimate can be made on site using techniques of simple measurement with manual analysers and test-kits, enabling a rough semi-quantitative overall determination of the following in a matter of minutes:

— presence or absence of a parameter (e.g. test for presence of hydrocarbons);

— indication or detection of a family of substances without distinction of components (e.g. colorimetric indication of heavy metals, nitrates or phenols);

— simple physicochemical measurements (e.g. pH or conductivity);

— and so forth.

On the other hand, procedures which are most onerous in terms of time, technical competence and expense, require specialised efforts for attaining access to the surroundings to be characterised and sampling according to precise procedures. These samples will then be sent to a laboratory specialising in sophisticated methods for, say, identification of the various molecules present, their concentration in the medium under study and the state in which they are found.

Between these two extremes of investigative intensity, all kinds of combinations are theoretically possible, each stipulating that the investigator shall combine efficacy and accuracy of techniques with expeditious execution at the lowest possible cost—an impossible task indeed!

These thoughts, relatively new in France, correspond to an awareness that could only have developed under the following pressures:

— the latest laws concerning management of water resources, discharge of effluents, elimination and disposal of wastes and protection of the environment;

— European directives and internationalisation of markets which tend to globalise the technical procedures and regulations for handling such problems.

The foregoing prompts laying down detailed guidelines for environmental diagnosis. However, we shall restrict ourselves to a brief review of some of the main features—and relative limitations of the following:

— in-situ monitorings,

— sampling plans,

— laboratory analyses,

— presentation and analysis of results.

3.5.2.1 *In-situ Monitoring*

Measurements conducted directly on site usually have two advantages: they are fast (results obtained immediately) and often inexpensive (compared to the more traditional approach of sampling and analysis in a specialised laboratory).

However, they have significant limitations: often they are not precise and may provide only an overall semi-quantitative or indirect result, necessitating subsequent interpretation and may turn out to be only an 'indication' rather than measurement of a direct parameter of pollution. Thus the data from a battery of georadars will be interpreted in terms of probability of finding metal objects buried in the subsoil, the exact nature of which can only be confirmed by direct surveys. Or again, the measurement of petrol vapours present in a subsoil will only indicate the existence of hydrocarbon pollution under the surface at that location.

A margin of some amount of uncertainty thus persists in case of results obtained by on-site measurements. It may be noted that if this margin is to be brought down, the complexity of measurement—and thus the cost and delay in results—increases and one perforce resorts to methods of measurements traditionally used in the laboratory.

The objective of the various measurements made on a contaminated site may also not be the same.

Generally, and as a first step, measurements made in situ seek to target the contaminated zone (zones) as rapidly as possible and at minimal expense. They enable crude delimitation in surface area of the extent of contamination in various recognised media, without necessarily being very accurate about the nature of the components themselves, their actual concentrations, and the state in which they are found. For example (Fig. 3.5), a plan for measurements of gas in the soil or monitoring of borings for water would permit delimitation of the surface area of a zone contaminated by hydrocarbons with an oil-film floating on the phreatic surface; this would determine the manner in which sampling of various media and laboratory analysis of the samples should be planned.

Measurements on site at this first step may be chemical, physical or biological. Table 3.2 lists some measurements that can be carried out on site in the framework of an environmental diagnosis, with their respective objectives and characteristics.

At the second step, measurements on site can be determinant for characterising the hydrodynamics of the water table. Being a medium perpetually in motion, the water table is not very suitable for monitoring through samples taken, stored and transported to laboratories away from site. Not limited only to investigations of a polluted site (they are also used in all explorations for water resources for example), these measurements are mainly concerned with:

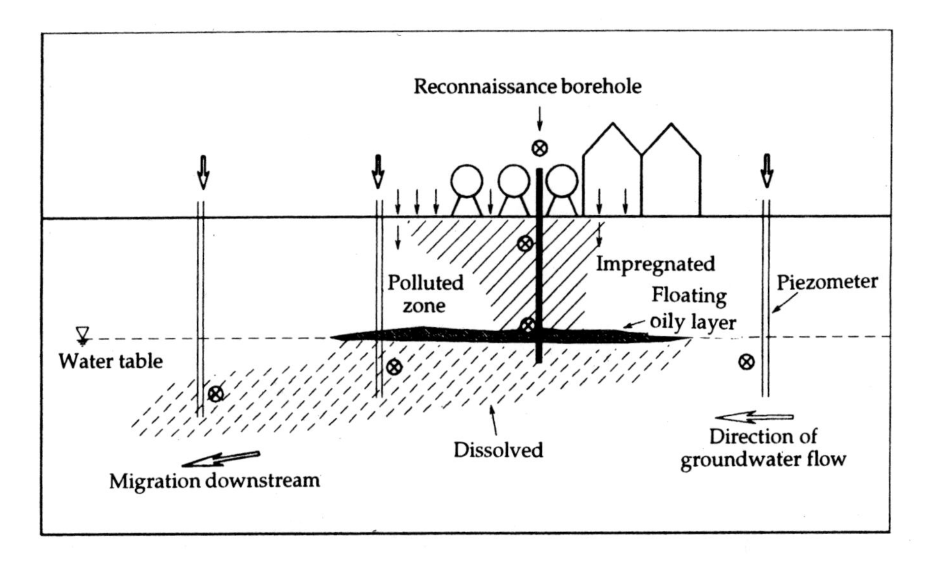

Fig. 3.5: Scheme of measurements on site in an environmental diagnosis (crosses indicate place of measurement in situ for physical parameters and arrows that for chemical parameters).

— variation in piezometric levels (depth of water in a well),
— permeability of aquifer strata,
— pumping tests which enable measurement of lowering of the water table due to pumping and hence the influence of water withdrawal at a certain rate on the phreatic surface.

These various parameters make it possible to define the direction and velocity of the groundwater flow, its discharge etc. These characteristics are essential if one wants to estimate how a pollutant is going to spread or migrate in an aquifer and also under what conditions it would be possible to install operations of in-situ remediation (see Chap. 5).

3.5.2.2 Programme of Sampling

A programme of sampling ought to enable obtaining information as complete and accurate as possible, with minimal equipment and time. This, together with the information collected during the course of survey of documents and measurements on site, constitutes the factual basis of the diagnosis (Table 3.3).

Maintaining a balance between the minimum necessary information and the means used for obtaining it is a difficult exercise; no 'miracle recipe' exists since each situation is specific. Although some overall basic rules have been established for preparation of the plan and mode of sam-

Table 3.2: Main environmental measurements to be carried out on site (sample table)

Type of apparatus/ measurement technique	Measurement	Media/component of environment on which carried out
Colorimetric/spectrophotometric test	Concentrations of specific elements (metals, anions, etc.) or chemical compounds (phenols, hydrocarbons)	Underground or surface water (sometimes, solid wastes or soil)
Immuno-enzymatic test	Quantum of organic elements (PAH, PCB etc.)	Water (sometimes, solid material)
Gas—chromatography	Identification of organic molecules constituents	Soil-gas, water
Photo-ionisation (infra-red spectrometry)	Index grouping the ensemble of volatile organic elements	Soil-gas
Colorimetric indicator tubes	Estimation of quantum of a specific volatile element	Soil-gas
pH meter, conductivity meter, thermistor etc.	Physiochemical characteristics	Underground or surface water
Level probes	Depth of phreatic layer, thickness of floating pollutant layers	Underground water (in boring)
Fluorescent lamps UV	Detection of hydrocarbons	Soil, solid wastes

pling, nonetheless it is always necessary to adapt them to the particular conditions of the study. Two extreme situations exist, one consisting of mere collection of a 'handful of top-soil' during a short visit to the site and the other implementing a lengthy scheme of drilling, densely distributed over the entire area to be reconnoitred.

To chalk out a plan of sampling it is necessary to answer the following three questions:

— What are the goals to be attained?
— What will be the order of sampling and the size of samples?
— Which techniques and mode of sampling will be used?

Plan of sampling

The plan of sampling depends on site logistics and soil features. Its scale and size are directly related to the degree of prospection envisaged. It is planned with due consideration of the stated goals and the amount of information pertaining to the surroundings of the site acquired beforehand, (pedological, geological and hydrological aspects) as well as the potential contaminants (nature of pollutants, quantities, sources etc.).

• In the course of designing the **plan of investigation** of soil and subsoil, the plan of sampling should be such as to avoid overexpenditure due

Table 3.3: Flow chart of diagnosis

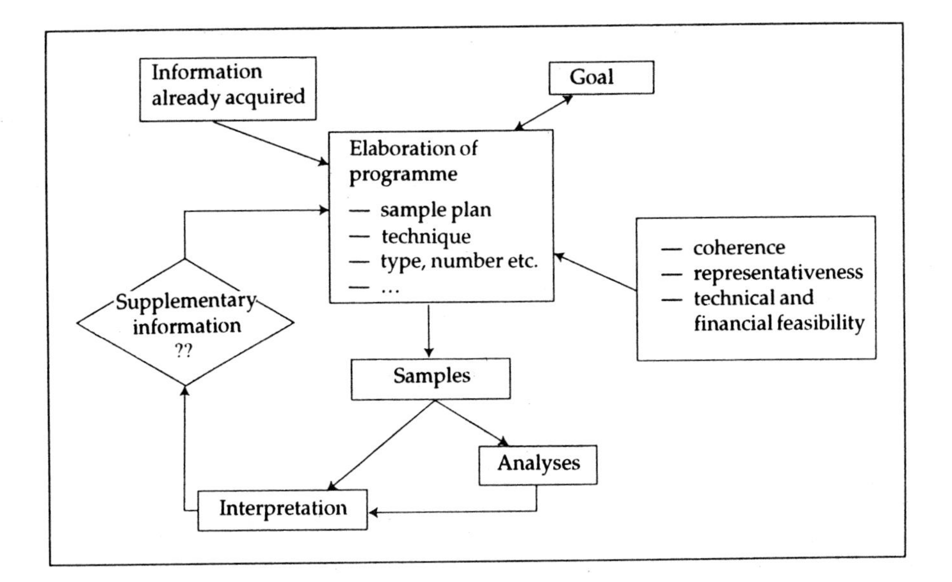

to too large sample size but at the same time omit no potentially 'interesting' zone.

Figure 3.6 (Keith, 1990) illustrates different types of sampling plans corresponding to the spirit with which the person in charge of the investigation approaches the problem. Keith's example pertains to the water of a lake upstream of which effluents of a factory are discharged; however it can also be applied to the sampling plan for soils of an industrial site.

The three major modes of distribution of samples over a surface to be studied are:

— **Systematic sampling** carried out according to a uniform grid that covers the entire surface, a sample being taken from every node of the grid or from the centre of each mesh. The shape of the grid (rectangular, circular or triangular) as well as the spacing of meshes (square mesh, transversal along one direction, radii of a circle) may vary. This mode of sampling is generally used for determining the spread of a pollutant and the decrease in concentration with distance from a source.

— **A priori targeted sampling** in which the samples are concentrated in a small zone identified as the most critical, according to information collected earlier or based on experience elsewhere. It can be seen from Fig. 3.6 that the immediate proximity of effluent discharge constitutes the location of highest risk. On the other hand, with this type of plan of sampling, one may stay in the dark with respect to what happens elsewhere. It is not always easy to identify all the risk zones in an industrial site from the background survey with the same degree of accuracy and reliability.

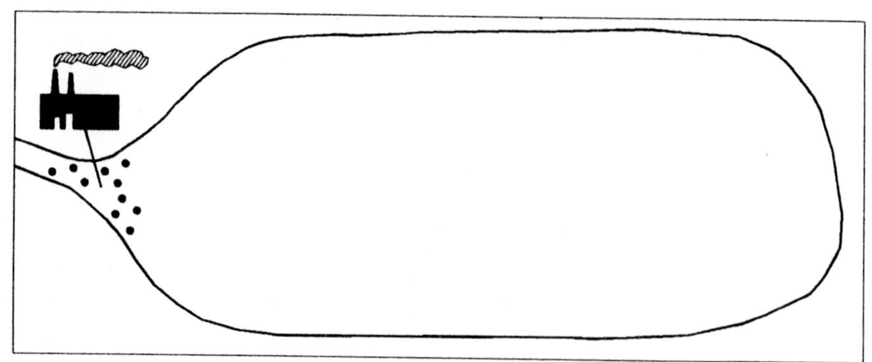

Targeted

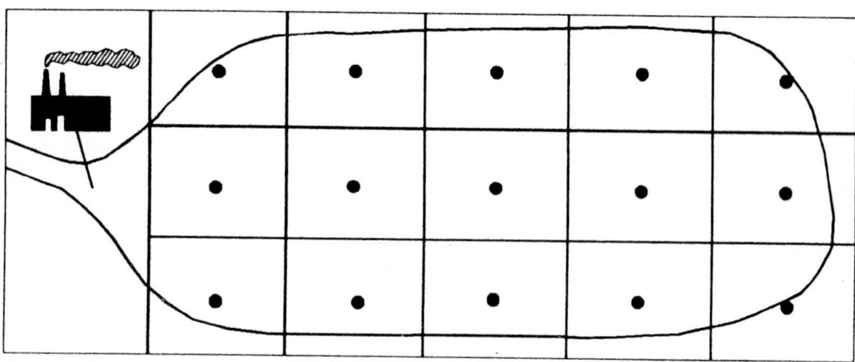

Systematic

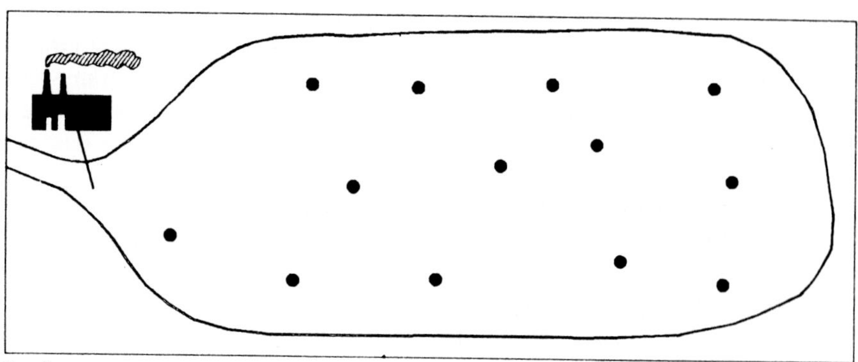

Random

Fig. 3.6: Types of sampling plans (Keith, 1990).

— **Random sampling,** implying that the sample collection is dispersed in an arbitrary manner over the entire surface with no predesign except perhaps a table of random numbers to fix the co-ordinates of the points of sampling. This type of plan corresponds to the minimum effort but suffers from the major disadvantage that actual characterisation of the site is difficult. Actually, random sampling is efficient only if the site is perfectly uniform which, in reality, is never true! Nevertheless this mode is subject to the least systematic bias.

In practice, one generally combines the three types, giving greater importance to the first (systematic) or the second (targeted) depending on what one is looking for.

Thus if the objective of the diagnosis is to carry out an observation of the state of locations, for example with a view to selling the site, one would prefer a more systematic approach in which all of the suspect zones are taken into account.

On the other hand, if the diagnosis is used for treatment and absorption of an accidental pollution, the sampling will be very targeted, focussing around the source of the accident and relatively neglecting other parts of the site.

• In contrast to a diagnosis carried out on soils, an **investigation of the groundwater** should take into account hydrological dynamics, tending on the one hand to shift downstream the area of dispersion of an eventual pollution and, on the other, to integrate the traces of geochemical history coming from the upstream with the characteristics of the site. Depending on the amount of information required, investigation of the groundwater will be more or less detailed and expensive.

There are always three distinct geographic zones to be prospected: the site itself and the regions upstream and downstream with respect to the groundwater flow (Fig. 3.7).

Analysis of the quality of the groundwater usually necessitates less work than investigation of the soils on the prospected site when the water is conveniently located, by taking into account:

— flow of groundwater (direction, slope...),
— surface spread of the pollution,
— behaviour of the potential pollutant in water medium,
— objectives of detection.

On the **upstream** one or more points of control are necessary for observing the quality of the water before its arrival on the site. If there are several potential sources of contamination of the water table upstream of the site under study, all these will have to be taken into account. Sampling of control points upstream is carried out in the existing wells; otherwise new borings have to be done.

On the **downstream**, to monitor the presence or absence of the pollution one or several sampling bores (already existing or installed during the plan of investigation) are needed. When the objective is the demarcation or extension of the affected zone and an accurate diagnosis or evaluation of the pollution, a larger number of sample bores are located along a circular arc or along transversals covering the entire hydrological zones downstream of the site.

Techniques and modes of sampling: Sampling holes

Sampling of solid media (soil, subsoil) and underground liquids can rarely be done directly; in almost all cases it necessitates creation of holes, bores or pits which allow access to the medium to be sampled.

To carry out investigation of the subsoil, the type of sampling hole chosen depends on numerous parameters, of which the major ones are:

— disposition of the site and its surroundings (surface or underground),
— type of pollution,
— type of terrain to be covered by investigation,
— maximum depth of investigation,
— presence and depth of the aquifer stratum (or strata—they can be superposed!) etc.

There are numerous techniques but the following are the most widely used at present (see photographs):

— **Pits or trenches** made by mechanical shovels (1 to 5 m deep, average depth 2 m). Easy and not very complicated, they allow good observation of localised terrain; however, they are limited in depth for reasons of expense and safety.

— **Borings done by auger or hand gouge**. Needing no machinery, this can be specifically implemented in the case of sufficiently friable terrain and shallow boring depths (in practice 0.2 to 1 m, may be up to 2 m); the amount of earth removed is small and the sample can be collected directly from the material falling to the bottom of the pit or attached along the wall of the auger during ascent.

— **Mechanical drilling**. Used particularly for investigations at greater depth (several tens of metres) and when the terrain is very hard, this offers a wide choice of possibilities and technical options with reference to level of borings as well as drilling machines. Being bulky, the machines take up considerable space on the site during execution of the work and are comparatively (much) more onerous than pits or manual borings. They range from the simplest motorised auger to a carrot drill with samples get enclosed in a sheath. Drilling is usually vertical but may also be inclined or horizontal.

— **Investigation of groundwater** is generally conducted with piezometers, designed and installed specifically for this purpose. A piezometer is attached to a mechanical drill that bores down to the water table (Fig. 3.8); the piezometer is then positioned at the depth at which water sampling is to be done.

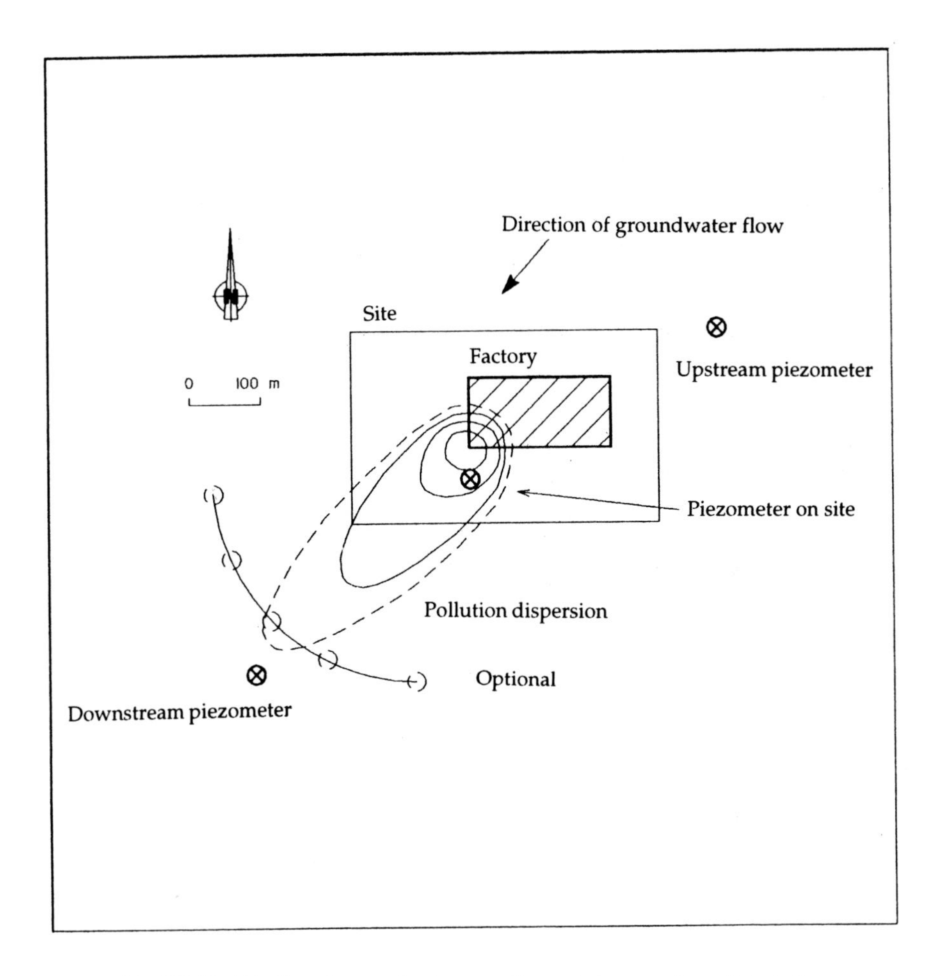

Fig. 3.7: Distribution of groundwater sampling-points around a polluted site.

Depending on the type of pollution, the work may be limited to the first few metres of the groundwater, e.g. in case of suspicion of floating hydrocarbons, or may extend to the bottom of the aquifer, e.g. when products not very soluble in and denser than water (say, halogen solvents)—or very soluble (nitrates) are apprehended. The piezometers are stationed at the depth to be reconnoitered, either at the water table surface, or deeper, or at various levels throughout the aquifer. Depending on the anisotropy of the medium, its heterogeneity and the type of pollutants, special equipment have sometimes to be used. Thus a sensor on the surface and another at the bottom of the aquifer, the two isolated from each other, allow distinct characterisation of the quality of water at all depths without

Photo no. 1: Sampling on sites: different types of tools and sampling holes (source BRGM). a) carrot drill; b) mechanical trench; c) motorised auger; d) pit dug by mechanical shovel.

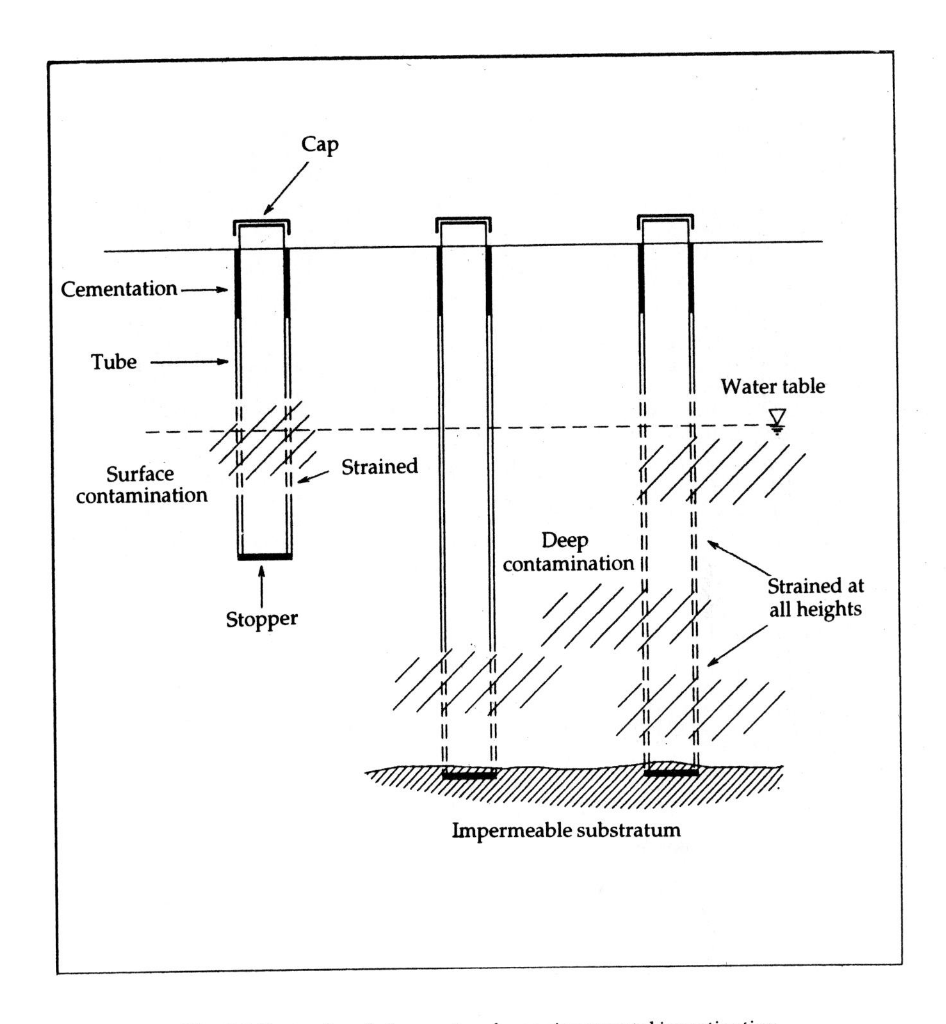

Fig. 3.8: Examples of piezometers for environmental investigation.

introducing the mixing factor. Isolating two sensors in the same bore is, however, delicate and it is often better to do two separate borings.

Less often, sampling of groundwater is done utilising other types of holes, for example an artisan well or a natural source or holes created for investigation of the subsoil (pit, trench...). If the existing wells and water-holes are relatively few (or not accessible), the holes created for soil investigation alone may not be very representative of the groundwater and would characterise only the locales where they are found. Relying totally on soil test-holes for sampling the groundwater is thus to be avoided, except when no other possibility exists.

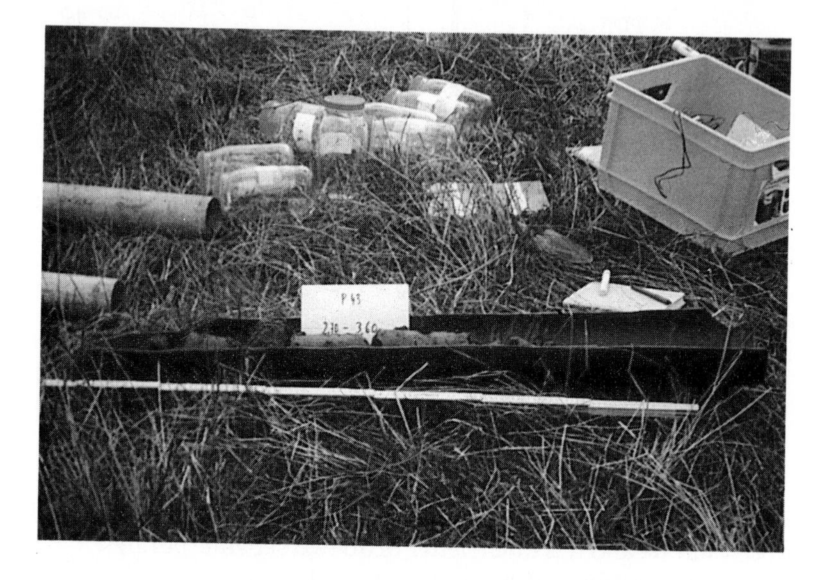

Photo no. 2: Sampling of polluted soil (source BRGM).

Techniques and modes of sampling: Drawing samples

Sampling the soil is usually done by means of a hand shovel; the quantity removed is variable but of the order of 1 to 2 kg. If the medium is heterogeneous, the quantity has to be increased to ensure the sample being more representative (several kg to several tens of kg, or even more). However to suppose that the samples can be taken directly to the laboratory for analysis is wishful thinking; they must be homogenised and divided into successive lots according to well-defined procedures until the quantity to be sent to the laboratory is obtained. The samples collected may be point-samples and thus representative only of the precise locale of sampling in three dimensions. On the other hand, samples may constitute a mixture of material evenly removed from a pit-facing or slope (the term 'vertical unevenness' is used) or along a carrot-boring; in this case the sample encompasses the characteristics of a certain thickness of terrain at a given locale.

For groundwater the techniques of sampling are multiple; we list three major modes employed directly in the piezometer:

— sampling tubes,
— syringes,
— pumps.

The sampling tube may be open at the top—used in this case for taking samples from the water table—with a non-return valve closed at the top (Fig. 3.9) or open at both ends. The tube is lowered to the desired depth, filled with water and then raised, a ballast being sent along the cable for closing the two openings, which ensures retention of water in the tube.

Syringes work on the same principle as a medical syringe. The apparatus is connected to a surface pump which activates a piston for filling the 'receiver' when the desired depth for sampling is reached (Fig. 3.9).

Pumps are termed 'surface' or 'immersed'; in the first case water is sucked from a certain depth to the surface—in practice these pumps are effective for depths not exceeding 7 to 8 m. On the other hand, the immersed pump is lowered in a well to the desired depth without limit and water is made to flow towards the surface.

In some cases water can be collected directly in bottles, at the outlet of a pump for example, in a pit or in a well of large cross-section.

A prior evacuation of the piezometer should be done by means of a surface or immersed pump before sampling. This practice ensures elimination of water which remains in the bore and assurance of a more representative sample of groundwater. Experience has shown that it is necessary to pump a volume of water equal to 3 to 10 times the volume of the piezometer, to obtain clear water, in equilibrium with the groundwater. The degree of turbidity of pumped water is actually inversely proportional to the representativeness of the sample. Figure 3.10 illustrates the variation of physiochemical parameters of water relative to pumping (Puls et al., 1991); in this particular case one observes that an equilibrium is attained after a volume equal to that of the tubes has been pumped.

In some cases surface water (river, marshy zone, source...) has also to be sampled, especially when it is in direct contact with the groundwater. Sampling is then done by means of bottles.

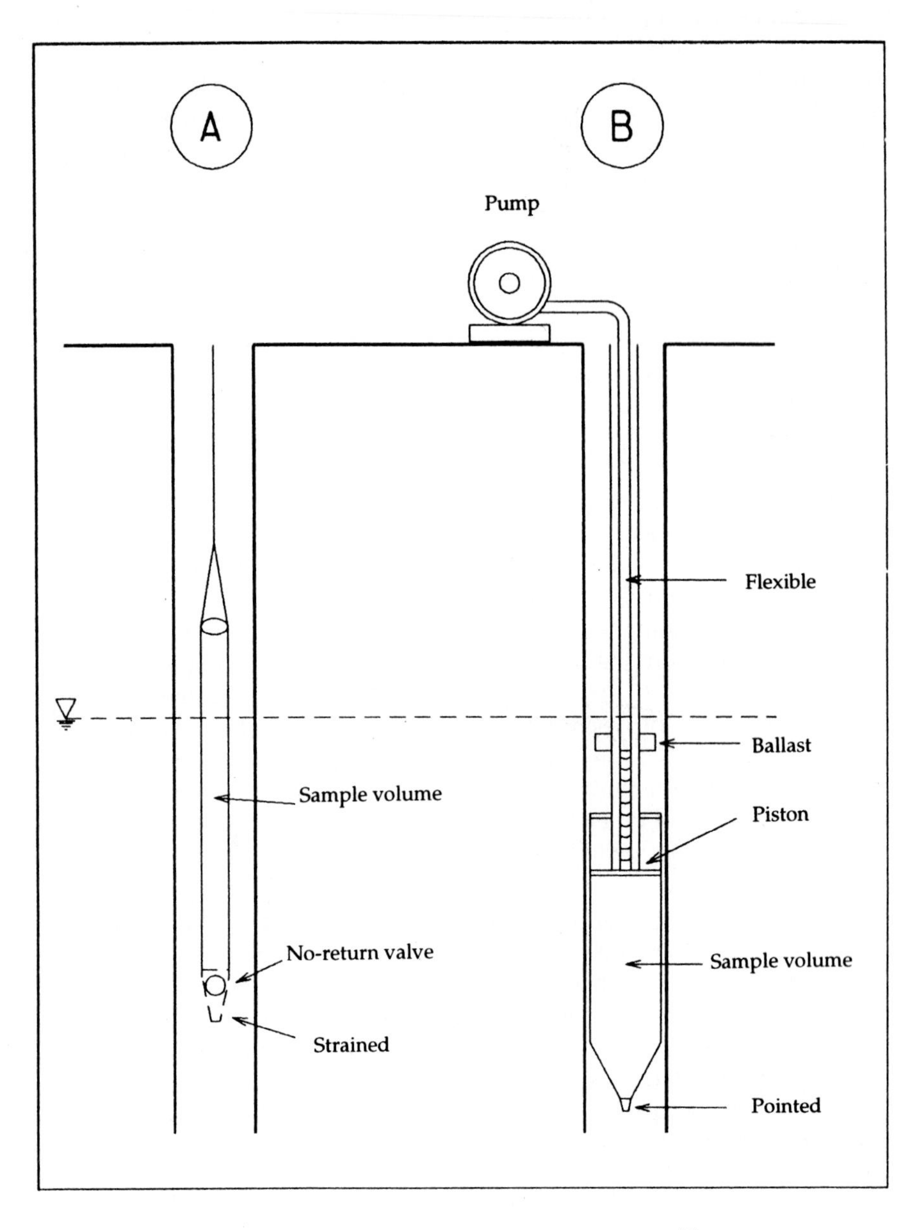

Fig. 3.9: Examples of water lifter (A) and syringe (B).

• **Packaging samples**

Samples of solid materials are mostly packed in hermetically sealed glass jars and eventually in plastic bags; packing in glass is obligatory for organic products but plastic is permissible for metals and mineral

Photo no. 3: Sampling of groundwater by means of a pump (source ANTEA).

products. Sometimes aluminium foil is used in closing the jar to preclude absorption of certain contaminants on the rubber seal (especially PCB and chloro-solvents). Finally, in taking out samples of soil for analysis for volatile organic compounds some operators recommend keeping the col-

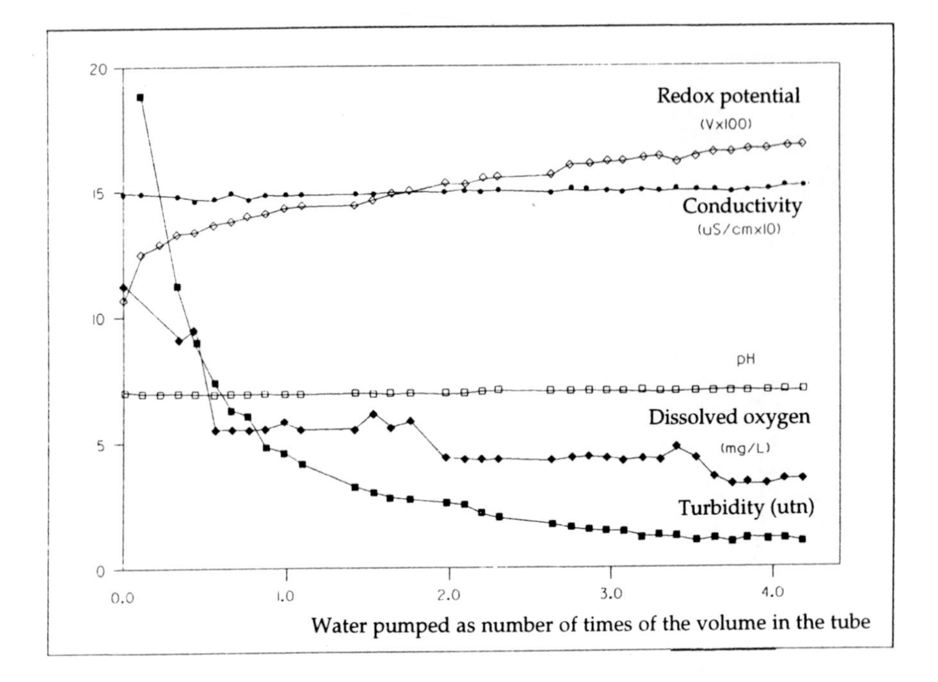

Fig. 3.10: Variation of physicochemical parameters of sampled water as a function of the volume pumped (after Puls et al., 1991).

lected soil samples in a hermetic sheath, which should be opened only under a hood at the time of analysis, to avoid loss of the organics.

Water samples are put directly into suitable bottles, mostly selected by the laboratories in which they are to be analysed. Depending on the type of substance to be measured, a specific type of bottle is used, certain precautions are taken and specific methods of preparation observed during filling.

For example, for total hydrocarbons the water to be analysed is put into a glass bottle of one litre after acidifying it with sulphuric acid; for anions (nitrates, phosphates, chlorides etc.) water is filtered and put into a polythene bottle of 250 ml and so on for various contaminants.

• **Precautions during sampling**

Sampling of water and soil should be done simultaneously so as not to interfere (or only minimally) with the environment and not to alter the nature and composition of the samples. Certain precautions must be taken during sampling; some are systematic and have a tendency to be generalised; others are more specific and constitute part of the know-how of sampling. In any case the 'level of precaution' attained remains difficult to judge.

The precautions commonly taken are:

— use of hand gloves and changing them after each sampling;
— cleaning of tools (shovel, hammer etc.) after each sampling;
— deposition of earth on plastic sheets, changed for every pit and boring;
— transfer of samples into an icebox at low temperature (around 4°C) and storage in a refrigerator or, in the case of solid samples, in a freezer.

Cleaning of water drills and piezometers with acid or tripolyphosphates is prohibited. So is injection of compressed air ('airlift'), especially if the presence of volatile organic products is suspected. Only pumping is recommended for cleaning and 'development' of bores or drill-holes.

Furthermore, it is always advisable to take duplicate samples, either over the totality of samples or at least partially (for example 10% of the samples, selected at random) as a check on the reliability of analysis (these duplicates are analysed in a laboratory other than the one in which the larger ensemble is analysed).

Finally, it must be emphasised that the design of the sampling hole-network should never augment spread of an existing pollution. Also, all pits and trenches must be carefully closed and the piezometers capped.

3.5.2.3 Laboratory Analysis

Should the analysis be carried out on site or in a laboratory? This is an important question in the investigation for the following reasons.

— It constitutes an objective input in decision-making during diagnosis; in most cases it is the results obtained during analysis which lead to determination of the reality of an eventual contamination.

— It is essential because it generates information that cannot be obtained by other means.

— On the other hand, it involves a long and complex process in which elementary errors accumulate at all levels of intervention and in some cases with irreversible consequences for diagnostic conclusions.

Analyses carried out in specialised laboratories are grouped by family of substances, organic in one family and mineral in the other. The analysis for one family enables obtaining results for a series of compounds belonging to that family, in general the most common and those most 'supervised' under government regulations. If the substance one is looking for is not included in the list, a special request must be made. Table 3.4 presents the ensemble of groups of substances currently of greatest concern.

In addition, for water samples certain parameters not corresponding to the group of specific substances nonetheless enable characterisation of the state of the quality of the medium. Prominent among such parameters are pH (measure of acidity–alkalinity), conductivity (hardness of water), suspended solids or SS (level of turbidity), chemical oxygen demand or COD and 5-days' biological oxygen demand or BOD_5. The last two

Table 3.4: List of families of substances usually analysed

Family	Detection of	Examples of method
Organics		
Total hydrocarbons	alcanes, alcenes	Spectrophotometry, infrared after extraction in CCl_4
Phenols	presence of group	Colorimetry after distillation
Monocyclic aromatic compounds (volatile)	benzene, toluene, xylenes etc.	Gas-chromatography with ITD after 'purge and trap'
Polycyclic aromatic compounds	Fluoranthene, pyrene etc.	Chromatography in liquid phase (HPLC) after extraction with dichloromethane
Polychlorobiphenyls	206 congeneres	Gas-chromatography with ECD detector after extraction with hexane
Organochlorine and organophosphorous pesticides	aldrine, hexachlorobenzene, DDT, malathion	Gas-chromatography with ECD or thermionic detector
Polar solvents	cetones, alcohols	Gas-chromatography with FID detector after 'head space' extraction
Chloro-solvents	trichloroethylene, dichlorobenzene	Gas-chromatography after 'purge and trap'
Pthalates	plastifiers	— same — PCB
Herbicides	atrazine etc.	— same — pesticides
Amines	pyridine, aniline etc.	Colorimetry after acidic extraction
Minerals		
Heavy metals	Pb, Zn, As, Cd etc.	Atomic absorption spectrophotometry after acidic extraction
Cyanides	Total CN, free CN	Colorimetry after decomposition and entrainment
Halogen anions	Cl^-, Br^-, I^- etc.	Ion-chromatography after lixiviation and passage over resins
Anions	NO_3^-, PO_4^-, $SO_4^=$ etc.	— same —

parameters represent the state of organic contamination of the analysed water: the higher the requirement, the more contaminated the water.

One of the main difficulties in the analytical step is selection of the laboratory. Numerous criteria have to be taken into account at this stage to avoid the risk of invalidating the entire investigative processes carried out for a site, the degree of technical competence and sincerity of the laboratory being keys to a successful diagnosis.

Aside from criteria of costs, the time needed for delivery of results and respect for confidentiality—criteria of primary importance—considerations of quality and reliability of the laboratory are also essential.

In particular, the following questions have to be posed:

— What is the technical capability of the laboratory? For example, of the 132 substances particularly supervised at the European level, which are the ones whose analysis can be ensured?

— Which techniques can it implement and what thresholds of detection can it guarantee? Are the analytical methods used the standard methods (AFNOR, ISO, EPA etc.)?

— Does it possess references for the relevant studies? Can other clients be contacted?

— Does a programme of quality insurance (norm ISO 9001 etc.) exist? Is the laboratory approved for analysing potable water? Is it a member of the network of accreditation such as RNE for example?

— Does it accept responsibility for destruction of a contaminated sample after analysis, in strict accordance with legislation?

— Does it assure the conservation of samples during an established contractual period and in accordance with the clearly specified procedures?

— etc.

3.5.2.4 Presentation and Interpretation of Results

Presentation

In a diagnostic report presentation of data may differ markedly depending on the goals of the study as well as the equipment and methods employed in the operation.

Generally speaking, the raw data should be presented systematically, mostly in tabulated form, annexing the laboratory methods of analysis. One may also append documents of more elaborate interpretation, e.g. maps showing geographical distribution or temporal graphs of evolution of a parameter over time). Table 3.5 depicts the major forms of presentation of data, from raw results to sophisticated cartographic interpolation.

Procedures for in-situ measurements, sampling analysis, packaging and conditioning of samples, transport and storage of samples etc. should be presented systematically in the report (mostly as a technical annexure).

Interpretation

The primary objective of these various documents is, of course, presentation of data in a clear and unambiguous form; but they also serve to highlight the elements of interpretation that the person responsible for the study takes into account while finalising the diagnosis.

The last part of the work is certainly very important but also one of the most difficult to accomplish. Aside from knowledge of the techniques mentioned above, it implies definite experience in this type of investigation. The know-how of the expert governs, as a matter of fact, the decisions made are sometimes loaded with consequences in terms of risks for the neighbouring population, financial and technical investments by the proprietor of the site and the restraints to be exercised by the administration.

Synthesis of observations of terrain and results of measurements and analysis do not consist merely of compilation of data. **Interpretation gives the data proper meaning, establishes the presence/absence of a contamination and, if present, its type, level, distribution etc. In every case the judgement of the expert will determine the follow-up action on the data, ending potentially with the evaluation of dangers and the choice of schemes of treatment and rehabilitation of the contaminated zones.**

Table 3.5: Modes of presentation and restitution of results

Theme	Objectives	Type of document
Location	Topography, geography	IGN maps, cadastral plans
Geological context	Regional and local geology	☐ Geological maps, sections, profiles (relating to terrain)
Hydrogeological context	Flow and slope of groundwater table, hydraulic characteristics	☐ Hydrogeological maps, direction etc. of flow, piezometry, test pumping
Works under investigation	Management of operations, location and description of works	☐ Organisational chart, plans of installations, topographical lay-outs, pedological profiles
Contaminants	Reference data	☐ INRS data, background values, guidelines, norms etc.
	Raw data	☐ Values, pollutant- and sample-wise. Range of values, pollutant-wise. Sources, pollutant- and entity-wise (water from such sector, soil from such pit). Statistical treatment and analyses.
Analytical results	Copies of laboratory reports/tables	☐ Concentrations of different elements and compounds, statistical analyses, variation with depth, in different media of environment
	Graphs	☐ Variation of concentrations over time. Profile of vertical and lateral dispersions
	Interpretational maps	☐ Isotenors. Extrapolations. Delimitation of suspect zones.

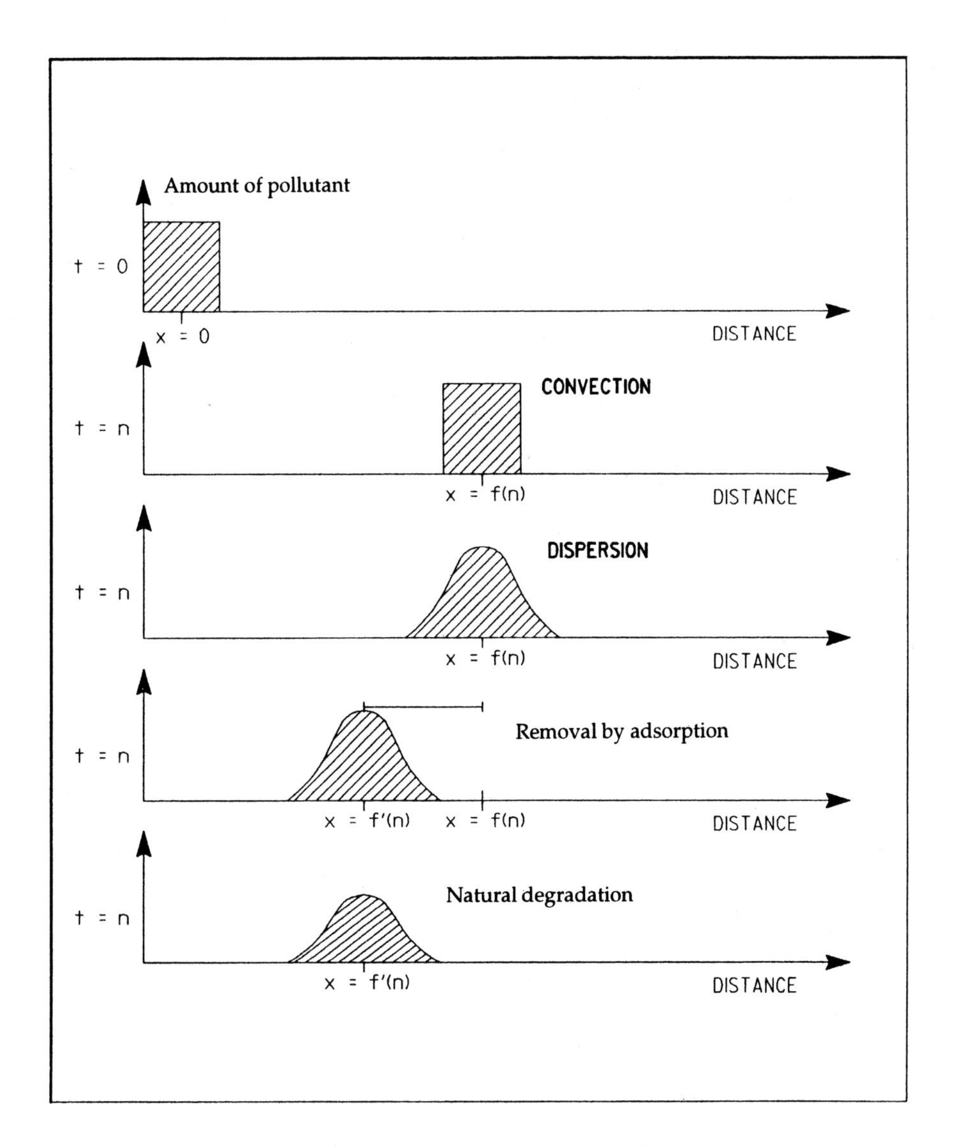

Fig. 4.1: Scheme of dispersion of a contaminant in a medium as a function of the transformation process taking place.

On the site itself the risk concerns almost exclusively the direct impacts on the persons working there (workers, suppliers, visitors etc.). Protection and safety of these persons constitute the top priority while impacts related to the ecosystem may often be of secondary consideration.

In case of the immediate environs of the site, the effect on the health of the population and on the natural environment are to be considered simultaneously. The immediate effects on the neighbouring population caused by site activity and the pollution generated by it are particularly critical in urban or suburban zones. As for the impact on environment, any protected zones, natural reserves or specific biotopes when present in close proximity will generally be considered first when planning specific and/or stringent measures of protection.

In addition to the immediate environment and the neighbouring population, the eventual consequences on the more distant population and environment have also to be taken into account; thus the ensemble of inhabitants drawing potable water from a tank downstream of the site may suffer the consequences of pollution of the ground or surface waters and inhabited zones situated sometimes at several tens of kilometres—or even suffer the ill effects of air pollution and fallout from factory emissions.

In conclusion, let us emphasise the 'domino effect' which pollutions can create in regions 'hydraulically' downstream or downwind of the polluted site due to propagation of contaminants.

4.3 ATTEMPT TO TYPIFY RISKS AND DANGERS

4.3.1 Risks for the Target

The risk represented by a contaminant that reaches its target can be identified in many ways. Table 4.1 shows that the potential dangers to which the population may be subjected will depend on the use to which a site is put or the activities existing in its proximity.

Table 4.1: Examples of potential dangers as a function of activities executed on or around the site

Usage		Potential dangers to a population
Agriculture for food materials	→	Mobilisation and transformation of contaminant(s) by plants
Domestic gardens	→	Toxic effects due to direct contact with the soil
	→	Phytotoxicity
Exploitation of groundwater for human consumption	→	Contamination of water sources Toxic effects due to direct contact with water
	→	Ingestion
Domestic gardens, playgrounds, stadia	→	Direct contacts
	→	Corrosion and degradation of underground structures (sewers, storage tanks etc.)
Buildings (storage, production, commerce)	→	Release of vapours, inhalation of released vapours, contamination of drainage networks
Underground structure (storage, transport)	→	Fire, explosions

Depending on the contact with the contaminant, the mode of exposure to which humans are subjected represents a deciding factor in the type of risk run. Thus we distinguish between:

— direct dermal contact, especially for personnel involved in the site activity and handling of the contaminating substances;

— inhalation of gases or particulates (dusts, aerosols) affecting not only the site personnel but also the neighbouring population and sometimes distant inhabitants as well (long-distance fallouts from emissions);

— direct ingestion of contaminant or contaminated soil, of relatively rare occurrence; precautions have to be exercised vis-à-vis young children playing in the proximity of industrial sites;

— ingestion of water reserved for human consumption. This is the case of pollution of potable water tanks when contaminants miscible in water have travelled into the aquifer, sometimes over several kilometres from upstream and generally with a large time-lag, often several years, after discharge of the contaminant in the environment;

— indirect or unintended intake into the alimentary canal whereby persons are poisoned by consuming animal or vegetal products that are contaminated (meat, fish, poultry, fruits, vegetables).

Although risks of explosion, fire etc. are always imminent, the more critical danger is that due to the toxic nature of the contaminants. Toxicity to human beings is considered separately from that vis-à-vis natural environment, also called ecotoxicity.

Ecotoxicity is measured in terms of the nefarious effects observed in plants and animals used as test organisms. Laboratory analysis is done for the effects of contaminating substances on each species tested by administering varying doses, to determine the threshold beyond which the nefarious effects start becoming noticeable (this 'maximum concentration for which harmful effects are too small to be differentiated' is called TLV, threshold limit value). It is obviously impractical to test all living beings in all biotopes; one usually restricts oneself to particularly representative species, such as certain fishes, daphnia, algae etc. One has to keep in mind that these simplifications have an element of arbitrariness and the data obtained often need to be corrected by using appropriate safety factors.

Toxicity to human beings should take into account several parameters, in particular the following.

— quantity of substance which is tolerable for an individual on a daily basis with due regard for his age, sex and weight;

— effect of accumulation in different tissues of the human body;

— mode of intake of the contaminant—through inhalation, dermal contact, direct ingestion or indirect entry into the alimentary canal;

— carcinogenic, mutagenic... character.

Such toxicological studies fall within the realm of specialists; the conclusions drawn from them are complex and difficult to apply in routine work.

4.3.2 Risks at Source

Parallel to the risks due to exposure of target populations one may consider the risk at the source itself depending on its nature and the industrial activities carried out there and thence attempt to define the sectors most exposed to the risk of contamination.

In France at the beginning of 1992, the Ministry of Environment created the Bureau d'Analyse des Risques et des Pollutions Industrielles (BARPI) with the important task of maintaining a census of the pollution accidents that have occurred in various industries in France and classifying them according to economic activity. This census is based on information from several sources at the national level (information gathered from DRIRE, civil defence...) as well as the international level (UNO, OCDE...), which is entered in a single database (ARIA) and regularly consulted. In 1992, an inventory of pollution accidents activity-wise (Mansot, 1993; Table 4.2) revealed that transport of dangerous materials by road or by train represented maximum risk, with more than 22% of the total accidents recorded during this activity. Derailment of wagons carrying petrol from Chavanay and Voulte an Ardèche was a typical example of such accidents and will be

Table 4.2: Activity-wise, comparison of number of pollution accidents that occurred in France during the past five years (Mansot, 1993)

Causal activity for more than 2% of accidents for which origin identified	Number of accidents as per census[(1)]	%age of total number of accidents
Transport of dangerous material (train, road)	659	22.6
Wholesale business and distribution	323	11.1
Food-agriculture	237	8.1
Chemical industry	236	8.1
Timber industry	215	7.4
Metal works	135	4.6
Agriculture (non-food)	119	4.1
Textiles	99	3.4
Garages, service-stations	94	3.2
Water tanks, distribution and treatment of water	69	2.4
Paper, cardboard	67	2.3
Petrol refineries	68	2.3
Transport of dangerous materials (pipelines)	58	2.0
Total	2379	81.6

[(1)]Excluding accidents of undetermined origin.

and is discussed in the following chapter. Next in order was the wholesale business and distribution, with 11.1% accidents, almost all of them pertaining to fire. Mention may also be made of the food-agriculture, chemical and timber industries, each of which accounted for 7 to 8% of the accidents recorded. On average, these five sectors were responsible for one out of every two accidents.

The statistics published in 1994 corroborate the above tendency since 24% of the pollution accidents were associated with transport.

In addition, depending on the nature of the accident, it has been observed that the dispersion of dangerous material in the environment represents more than half the cases recorded while fire accounts for slightly more than 40%.

4.4 METHODOLOGY OF ASSESSMENT OF HAZARDS

Different methods are employed in various countries (Germany, Canada, the USA) to evaluate hazards or risks associated with pollution. The purpose is to estimate the potential hazard represented by a site, either for relative ranking of individual sites of an ensemble or proposing a scheme for rehabilitation of a particular site.

4.4.1 Methods of Ranking of Sites

Although the methods of ranking sites (termed 'Hazard Ranking System') differ in conception and technique of application, they nevertheless follow a common strategy and have the same objective. In all cases they are governed by the same basic principles.

Thus it is always a matter of a pragmatic approach, permitting analysis—and comparison—of several sites according to criteria that have become well established over time.

The method should take into account all the threatened components of natural environment, which also serve as 'potential routes of transport of the contaminant'. The existing methods are unanimous on this point and divide the natural environment into surface water, groundwater, air and soil/subsoil; in certain cases 'direct contact' of the exposed subjects with the contaminant is added to this list.

As a standard practice, analysis considers the three key locations of impacts discussed in the preceding section:

— source,

— along the route of propagation,

— target.

An analysis can thus be characterised by a table with the four natural routes of transport of the contaminant as columns and the three key factors as rows (Table 4.3). Evaluation is often quantitative, measuring the poten-

Table 4.3: Evaluation of hazards: scheme of notation in terms of media and key factors

Key locations of impact	Component of environment or route of transport				Total
	Surface water	Groundwater	Soil/subsoil	Air	
At source	Score 1–1	Score 1–2	–	–	Score 1–
Along route of propagation	Score 2–1	–	–	–	Score 2–
At target	–	–	–	–	Score 3–
Total	Score –1	Score –2	Score –3	Score –4	Integrated score

tial dangers by calculating a score for each entry in the table, which depends on an ensemble of specific characteristics for each case. The scores are subsequently combined to obtain an integrated score characterising the entire site.

Depending on the methods, the integrated score is obtained by addition or multiplication or a combination of the two or even by a system of averages. Certain more qualitative methods do not use scores but distribute the sites into predefined classes of 'hazards'.

Each method has its own scale of reference for scoring, which is generally included in a 'user's manual'. The scales and scores are based on the past experience of the designer of the method. The scale of reference comprises the points to be attributed corresponding to each characteristic to be considered in the evaluation. For example, in the American method, HRS (Hazard Ranking System—EPA, 1990) a score between 25 and 3 is assigned depending on the proximity to a surface water source (flowing water, lake...); if a site is situated on a water-bank, it is given the maximum score of 25 (Fig. 4.2) while if it is more than 2800 m away, it receives the minimum score of 3. Similar relations between scores and 'data of terrain' may be established for each characteristic identified in the method; such relations are generally determined during the survey phase.

In some methods, a part of the interpretation is left to the discretion of the evaluator who determines the score within the range defined by the method on the basis of his own experience and personal perceptions.

When the integrated score has been obtained, it is compared with a table of values defining the state of the site in terms of 'dangerousness'. In the HRS system the integrated score is the root mean square of totals obtained for each environmental component and varies from 0 to 100, proportionate to the potential hazard. Should this score be higher than 28.5, the site is classified as one which needs to be rehabilitated.

Every step of evaluation must be transparent and controllable; at no moment should it become a 'black box'. Evaluation is done on the basis of available information; in some methods the extent of knowledge of the site

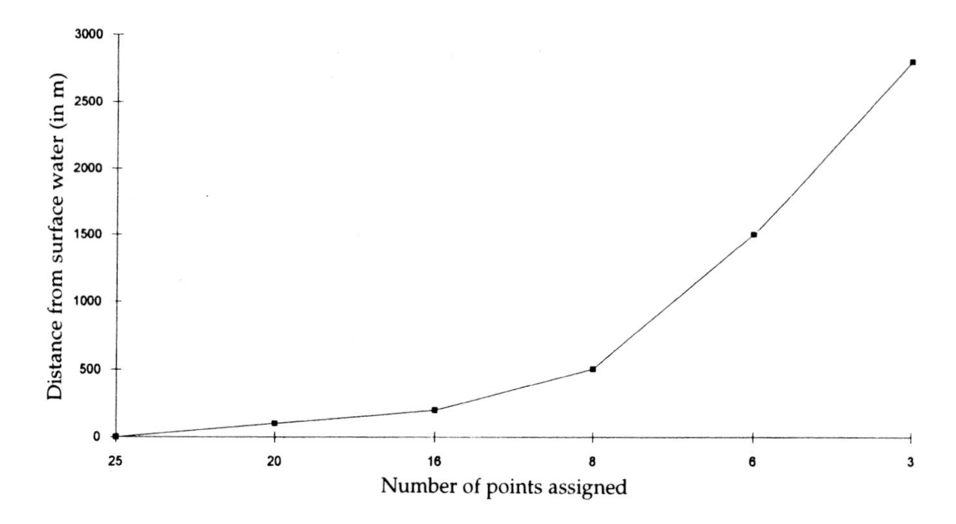

Fig. 4.2: Assignment of a score according to the HRS method (USA) for a given observed characteristic, here the 'distance between the site and the nearest body of surface water'.

carries a weightage with the scores defined. Since the process is dynamic by definition, it ought to be possible to recalculate scores afresh with the addition of new information. In this way the variation of the integrated score of a site and thus its dangerousness can be followed as one progresses towards a more and more refined diagnosis as also during remediation operations.

These methods are all relative but nevertheless enable adequate decisions in terms of responsibilities and solutions to rehabilitation of sites. Also they may or may not be comparable among themselves and may not give exactly similar results when applied to the same site.

A team of BRGM (Côme et al., 1993) evolved a programme of comparison of the ranking methods by applying them on test sites and/or by several users working independently. The conclusions of such studies are often surprising. As a matter of fact, employment of many methods for the same site and by the same user does not necessarily lead to concordant recommendations. A significant factor of subjectivity is associated with the user and analysis of the same site by the same method by several persons often results in divergent conclusions, implying different decisions. Finally, it is also clear that proper application of these methods requires vast descriptive data of the sites and very special training of the users. Côme et al. conclude that it is necessary to improve the methodology of evaluation of hazards by adapting it specifically to the legislative, economic and environmental contexts developed in a country. It may be noted that a similar

An example of comparative analysis carried out on three fictitious sites may illustrate the general principles discussed above and the nature of an evaluation of hazards.

Let the three sites possess the following characteristics:

1. Service station

Source: chronic leakage of petrol from the underground tank, large releases

Route of Propagation: slightly permeable soil, water table at small depth

Target: urban sector, danger of explosion

2. Factory for surface treatment

Source: effluents containing chloro-solvents and heavy metals

Route of Propagation: valley floor, proximity to flowing water, sandy terrain

Target: ecosystem of the downstream valley

3. Domestic wastes dump

Source: effluents containing organic matter, foul odour, eventual presence of unidentified toxic substances in limited quantities

Route of Propagation: clayey terrain, recharge area for groundwater

Target: rural zones and a storage tank of potable water 1 km downstream

Results of evaluation of hazards conducted for the three sites using the same method are presented in Fig. 4.3. In this case the integrated score is the sum of scores relating to the factors 'source', 'route' and 'target'. For site 1, the relative impact at the target is large and the overall impact is larger than for the other two. Corrective measures are definitely needed for this. For site 2, the source represents a major risk and a drastic plan of prevention should be put into action to protect the downstream natural environment. Site 3 has the lowest integrated score and presents the least hazard for the environment.

study conducted recently in the United States led to identical conclusions (Hushon et al., 1993).

New methods are currently being developed and validated at BRGM; they are being used in other professions and permit removal of the bias observed in the methods described above.

4.4.2 Methods of Hazard Evaluation

Methods for evaluating dangers (or hazard assessment methods) are applicable to a specific, well-characterised site and permit quantification of the 'absolute' impact or nuisance caused by a site. Here it is not a matter of assigning arbitrary scores, but of describing the migration of contaminants

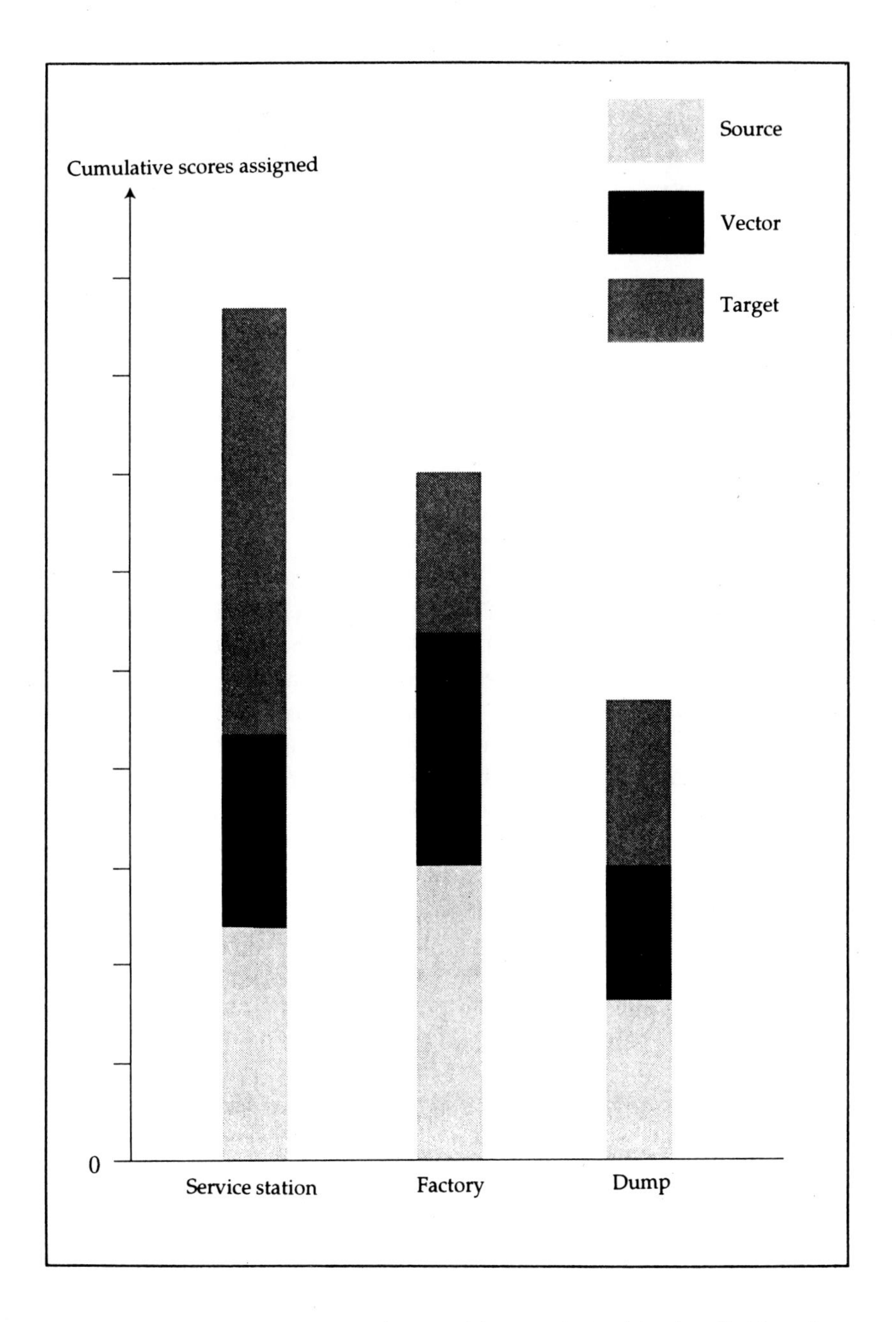

Fig. 4.3: Results of comparison of evaluation of dangers obtained for three fictitious sites.

from their source (for example, a defective dump) in a natural medium (such as groundwater) towards an identified target (the biosphere, a storage tank for potable water...). Use is made for complex computer-aided models. Such methods are used in the United States—code MEPAS (Hushon et al., 1993), or in the Netherlands—code HESP (ECETOC, 1990). Their use necessitates collection of vast data on the travel of contaminants in the natural environment and hence is reserved for a small number of special cases.

4.4.3 Strategy Developed in France

Depending on several factors, such as specific perception of environmental problems, economic and social conditions, level of technical know-how, culture etc., each country has formulated its own policy on the subject of management of polluted sites. Two well known examples of such strategy are: (1) the Dutch strategy based on the utilisation of concentration limits determined by national standards* which are often used by other countries too, and (2) the policy formulated in the USA by the federal agency for the protection of the environment (EPA) using the programme SUPERFUND. In France, the minister in charge of environment has opted for a policy of management of contaminated sites based on risk assessment. As a matter of fact, rather than using lists of standards which would anyway be arbitrary in view of disparities among situations in different regions (natural, geological, climatic, economic and social factors, state of urbanisation and industrialisation, administrative procedures and composition of wastes etc.) he has preferred to consider each case (each site) separately, thus subscribing to the continuity of the law pertaining to classified installations and to adapt the degree of intervention according to the level of actual risk. As a first step, each site is subjected to a simplified risk-analysis which enables classification of the sites into three categories: 1) sites needing detailed investigations in view of the gravity of the state of pollution before a possible curative intervention; 2) sites needing caution and vigilance; 3) ordinary sites with no environmental risks. This simplified evaluation of risks is based on existing information, if available, without insisting on a supplementary investigation. As a second step, the sites of the first category—and eventually those belonging to the second category—are subjected to an in-depth diagnostic and detailed evaluation of risks, necessitating major investigations and consideration of the ensemble of all risk factors.

---

*In this regard it appears that the Dutch policy is now proceeding towards a less drastic vision by also considering the 'risk assessment' aspect and correlating this with its direct standards-based approach.

4.5 COMPARISON WITH RESPECT TO THE CONCEPT OF NORMS

The methods employed for evaluation of hazard define, each in its own way, that level of risk beyond which some specific actions are proposed, such as additional survey, preventive and supervisory measures, and especially operations of remediation and rehabilitation of sites. As seen above, methods for the same site could lead to divergent results and a particular criterion could have a different weightage when taking a decision for one site vis-à-vis another. For example, the impact on the environment of a high concentration of heavy metals in a soil will vary according to whether the soil is permeable (highly sandy) or is formed of impermeable clay. Taking into account these measures leads to the question: Aren't we assigning two different weightages and hence obtaining two different decisions? As for a global environmental hazard potential at the scale of the planet Earth, shouldn't every pollution and every attack on the environment be considered harmful to humanity, no matter what the other circumstances are? Shouldn't it be considered that beyond a certain limit of concentration in the natural environment, corrective measures have to be adopted no matter what use the site is put to or how thin or dense the neighbouring population is?

4.5.1 Norms: Definition and Nature

This plan of protection of the environment is practised in some countries and usually with regard to some specific questions. It corresponds to laying down norms. These norms, generalised and accepted in some domains (discharge to the atmosphere, level of measured impacts, standards of quality for potable water for human consumption, surface-water-quality objectives etc.) are difficult to elaborate in the context of rehabilitation of sites, for soil and water.

In France there are no norms or guidelines which prescribe for each potentially contaminating substance the limit of concentration in the soil or in surface or groundwater beyond which remediation becomes a must.

In Europe the best known example is the Dutch 'guideline', which is often used as a reference in other countries also.

4.5.2 'Guideline' of the Netherlands

Elaborated in the second half of the 1970s (first edition in 1983), the 'guideline' is based on this principle: every site can be put to any use irrespective of the use to which it had previously been put. Starting with the fact that the country is highly industrialised but not large in area (about 33,000 sq. km) and that the natural conditions are fairly identical everywhere (sandy-clayey plains with the water table just a few metres below the surface), the Dutch planners have estimated that no site can be reserved for a specific use (industry etc.) determined by its level of contamination, even high, and it should always be possible to reutilise it for a different purpose (e.g. urbanisation, recreation...).

By definition the proposed system is referential, placing at the disposal of the authorities, industrialists and environment protection agencies, representative data enabling decision-making. In due course of time the authorities of the country may arrive at decisions with due regard for the norms and legal framework applicable in that country.

The referential system is established in the following manner. For each substance three levels of concentration are taken for water and the soil, corresponding to three levels of reference (A, B, C). Level A is considered a concentration perfectly normal and acceptable in the medium; it is akin to a 'background level'. Level B is an 'alert' level, beyond which it is advisable to conduct a detailed survey to eliminate or confirm the suspicion of pollution. Level C corresponds to non-acceptable concentration, equivalent to a serious threat for the environment, and corrective action needs to be taken immediately at such sites.

The Dutch experts maintain that these referentials do not constitute criteria for the quality of the soil or water; they also assert that the relevant lists are subject to revision in view of the acquisition of new knowledge of contaminants and evolution of technoeconomic criteria.

In 1993, a corrected version was prepared for submission to the Parliament. It includes only level A—concentration normally acceptable—and level C—the threshold not to be crossed. Between the two values lies the zone of uncertainty for which it is advisable to conduct supplementary investigations.

This referential was established on the basis of numerous scientific studies taking into account primarily the criteria of toxicity vis-à-vis human beings, ecotoxicity, mobility in the natural medium...vis-à-vis the various contaminating substances referenced in the guideline (van der Graag et al., 1991; van der Berg, 1992).

Threshold C, or maximum tolerance values, was particularly the object of numerous surveys.

The first version of the guideline was formulated on the basis of an ensemble of tests of toxicity on a large number of living species, animal and vegetal. For each species the 'maximum concentration corresponding to no observable effect' or NOEC (see above) was determined and the ensemble of these limiting concentrations determined for all the species tested, was analysed in terms of frequency (Fig. 4.4). Value C corresponded to the concentration up to which no adverse effects were observed in 50% of the species tested.

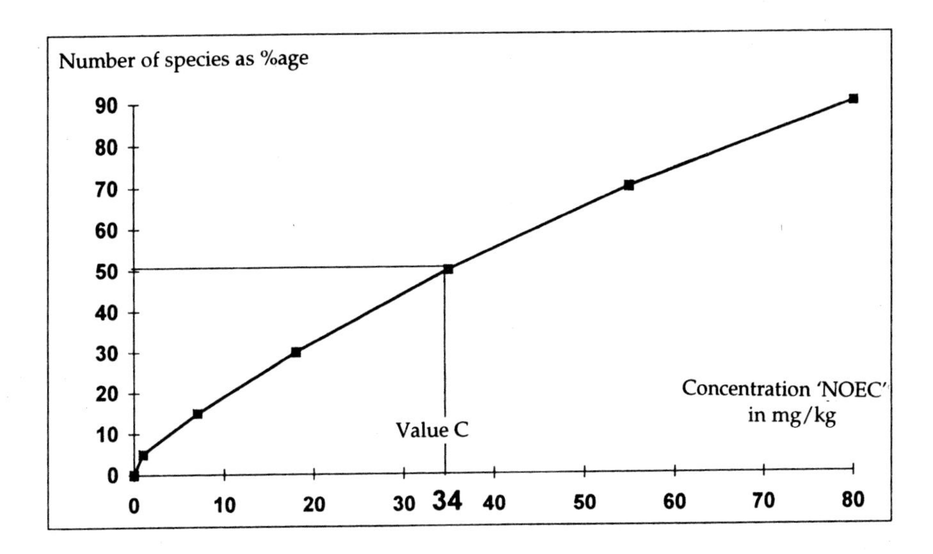

Fig. 4.4: Distribution of frequencies of 'maximum concentrations corresponding to no observable effect' (or NOEC) on all the species tested for a contaminating product.

In the second version of the guideline other parameters are taken into account for redefining and readjusting the values of C, with, among other things, toxicity for human beings and mobility of the contaminant in the medium. The number of parameters considered and their complexity render determination of such a threshold difficult and subject to numerous interpretations and approximations. Given this fact, one often finds divergences, sometimes large, among the referentials proposed by different organisations or countries (discussed later).

Threshold A, akin to a background level, is more easily determinable. Carried out in a more empirical manner, this determination has as its goal the proposal of sufficiently low values for constituting no nuisance for human beings, animals or plants and, concomitantly, preventing transfer of contamination from one medium to another, for example transfer between the soil and the groundwater. For water, it was often the norms of potability that were taken into consideration. For soils, one may consider the levels observed in natural soils away from all human activities. Although for organic substances the proposed values may be acceptable everywhere, the same cannot be said for mineral compounds, and more particularly heavy metals.

Regarding the latter, the form in which the mineral compound is found is critical for preventing all nuisance and mobility in the medium. In tropical zones some laterites contain accumulated heavy metals, such as nickel, cobalt, chromium, vanadium, or even arsenic and molybdenum which, when immobilised in the crystalline structure of oxides of iron and aluminium, are completely harmless for the population. Such metal concentrations may attain values of several hundred or even a thousand mg/kg, which is much higher than the values of A recommended by the Dutch guideline.

Furthermore, the natural background level of a mineral compound is always closely related to the geological context and the Dutch subsoil hardly represents the ensemble of the most frequent natural geochemical contexts. Thus in numerous regions of France the background levels of metals observed in soils on igneous or volcanic rocks are often much higher than the value A of the 'guideline', corresponding to no harm for the population. Table 4.4 compares for some heavy metals the thresholds A with the natural average abundance observed in sedimentary clays and carbonates (Turékian an Wedepohl, 1961) and the regional background levels determined in Lorrains during a study of soil sediments of the Orne Valley (Bonnefoy and Bourg, 1985). The differences are often very significant, whether positive or negative.

Table 4.4: Comparison of values A of the guideline with natural geochemical references in soils or sediments (all values in mg/kg)

	(Bonnefoy and Bourg, 1995)	Average natural abundance (Turékian and Wedepohl, 1961		Dutch 'Guideline'
Element	Regional background level	Clays	Carbonates	Threshold A
Co	16	74	0.7	10
Ni	138	90	11	100
Cr	39	225	30	50
Cu	20	250	30	50
Pb	35	80	9	50
Zn	133	165	35	200
As	43	–	–	20

4.5.3 *Examples and Comparison of Norms*

The Dutch guideline does not constitute an isolated example on the subject of elaboration of norms for rehabilitation of contaminated soils and water. As a matter of fact, several countries have formulated their own norms in this domain—in Europe, Germany and Great Britain, in North America, the USA and Canada, and also Australia. The list is not exhaustive.

Table 4.5 compares the 'limiting values' or norms acceptable for the contamination of soils and determined for some organic and common mineral substances in six countries, states or provinces: the Netherlands, Great Britain, Germany, Quebec, Utah and California (USA) and Australia. It

Table 4.5: Comparison of limiting values of some common contaminating substances

Contaminant (1)	Netherlands (3)	Great Britain (4)	Germany	Quebec	Utah (USA)	California (USA)	Australia
Lead	600	500	1000	1000	–	–	300
Cadmium	20	3	10	20	–	–	20
Arsenic	50	10	50	50	–	–	100
Chromium, total	800	600	250	800	–	–	–
Complex cyanides	500	1000	–	250	–	–	–
Phenols	10	200	–	10	–	–	0.5
Hydrocarbons total (2)	800	–	–	800	30–300	10–1000	–
Benzene	5	–	–	5	0.2–1	0.3–1	1
Benzo (a) pyrone	10	–	5	10	0.2–0.2	–	1
PAH total	200	500	–	200	–	–	20
PCB total	10	–	1	10	–	–	1

(1) in mg of contaminant per kg of soil
(2) total hydrocarbons constituting petroleum
(3) for these substances, old and new versions give identical values
(4) for these referentials, the most constrained values have been selected

illustrates the variations that may exist between one referential and another and the difficulty of formulating a norm coherent and valid for all.

From the data given in Table 4.5 we can make two observations:

— The Netherlands and Quebec are the only regions which have a complete list; also, they are identical except for one or two values.

— While the USA is particularly preoccupied with formulating norms for hydrocarbons constituting petroleum, the Europeans are more concerned with the problem of heavy metals.

Although some states, provinces (USA, Quebec) have followed an approach similar to the one on which the Dutch guideline is based, considering a level of reference by a substance independent of other criteria and evaluation of risks, others (Germany, Great Britain) have followed a different approach, introducing in their referential the notion of site usage. This implies that determination of thresholds of maximum tolerance (value C in the Dutch guideline) beyond which a corrective action becomes obligatory, will vary with the present or the future activity of the site. It is clear that the administration of these countries will be less strict, permitting higher concentrations of contaminating substances in the soil or water for sites reserved for industry, compared to sites for inhabitation, culture or recreation. Table 4.6 is an extract from a publication of Eickmann and Kloke (1991) elaborating the German strategy in this context. Similar to the Dutch 'guideline' these authors distinguish three categories of gradation in

Table 4.6: System of gradation of guiding values in terms of use to which site to be put (German referential, Eickmann and Kloke, 1991)

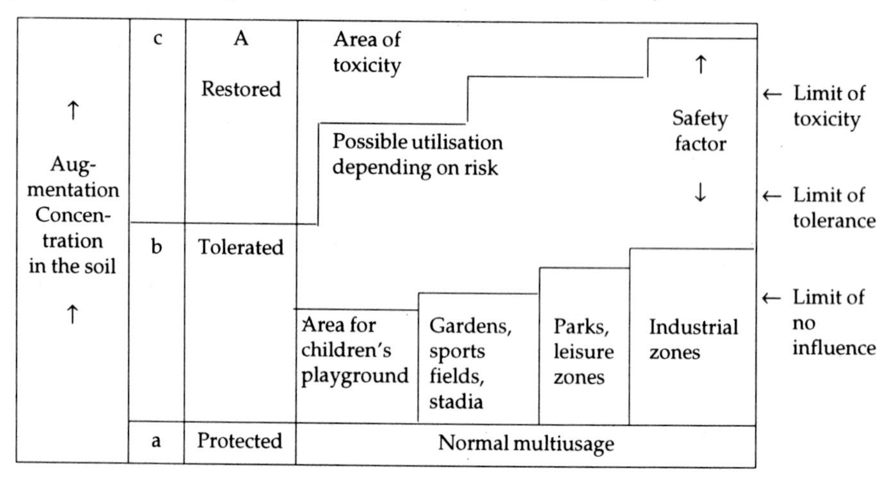

contamination: the domain of acceptance (A), domain of tolerance (B) and domain of toxicity (C). The thresholds permitting changeover from one domain to another are variable, however, depending on the type of use to which the site is to be put. Four types of usage are defined:

— areas for children's playground in which maximum strictness is exercised;

— areas for sports, pleasure zones and green belts;

— areas for traffic (road, train...) and parking;

— sites for industry in general (factory, workshop, warehouse...) for which tolerance in respect of the contaminating substance will be greater. The same system of gradation is formulated for agriculture, varying from commercial gardens to ornamental plants, with fruit cereals and industrial crop patterns successively in-between.

The threshold values limiting the domain of toxicity may have large variations: thus for arsenic in children's playground it is 50 mg/kg; for industrial sites with protected soil, 200 mg/kg; for PCB, 1 to 15 mg/kg and for benzo-pyrone (the group of polycyclic aromatic hydrocarbons), 5 to 10 mg/kg.

In France, in the absence of such a national referential, only two types of limiting values are used for soils and water: the norm of potability of water for human consumption and the highest concentration of heavy metals for the sludge or manure applied on agricultural soils. Both norms satisfy the limits prescribed by the European Community.

V

Remediation

In collaboration with Isabelle Le Hecho and Fabienne Marseille

5.1 SELECTION OF REMEDIATION PROCEDURE

After the diagnosis has been completed and the risk evaluated, before considering a plan of remediation, *one ought to decide whether or not remediation and rehabilitation actions are needed.*

This decision is generally made in consultation with the concerned administration, the proprietor of the site (and/or its user) and the agencies and experts dealing with technical aspects. The decision per se does not generally presume the techniques nor the actions that will ensue but rather profiles the environmental 'idyll' to be striven for throughout the process of rehabilitation.

After making a definite decision to remediate an entire site or a part thereof, one has to attend to the following questions: what procedure and action(s) should be proposed to attain the desired goal?, what technique(s) or methods of remediation should be selected and what will be the modalities for carrying out these operations (technologies, duration, cost...)? These considerations should take into account the technical, economic as well as social (in the broad sense) feasibility. From amongst the various possible solutions, the one that gives the best 'quality-to-cost' ratio should be selected.

Depending on the parameters, which sometimes overlap closely, the feasibility of a plan of remediation of a site will, by definition, be complex. Also, the proposal may entail examining several alternatives and considering a number of variables depending on the technological options envisioned as well as the technical and financial constraints relevant to the case (EPA, 1993a, b). In many cases resolving the problems will require not a unique solution, but a complex of sequential operations, each of which takes care of a specific 'part' of the overall rehabilitation.

Without going into the details of costs (see Chapter 6), experience has shown that an increase in the efficiency of an operation—or a series of sequential operations—of remediation varies exponentially with the investment made for carrying it out. Often the tangible results—sometimes spectacular—quickly realised on initiating action (Fig. 5.1), seem to fade out later in spite of no change in the rate of investment. Eventually one attains a plateau of output or efficiency, depending on the chemical and/or physical process, independent of the investment made. It may be relatively easy to start with a level of pollution at 100 and attain a level of 10 or 1, say, by pumping out a large quantity of hydrocarbons floating on the phreatic surface, or by excavating several cubic metres of soil contaminated with heavy metals, but it would be vastly more difficult to subsequently attain the background or desired level of 0.1 pollution, for example, by pumping and treating the groundwater for a number of years.

In view of these facts, one may reconsider the delicate problem of defining the objectives of remediation (Ricour and Lecomte, 1994). Cursorily defined at the start of the operation, the objectives would require balancing the requistioned level of rehabilitation with the standard of life of the popula-

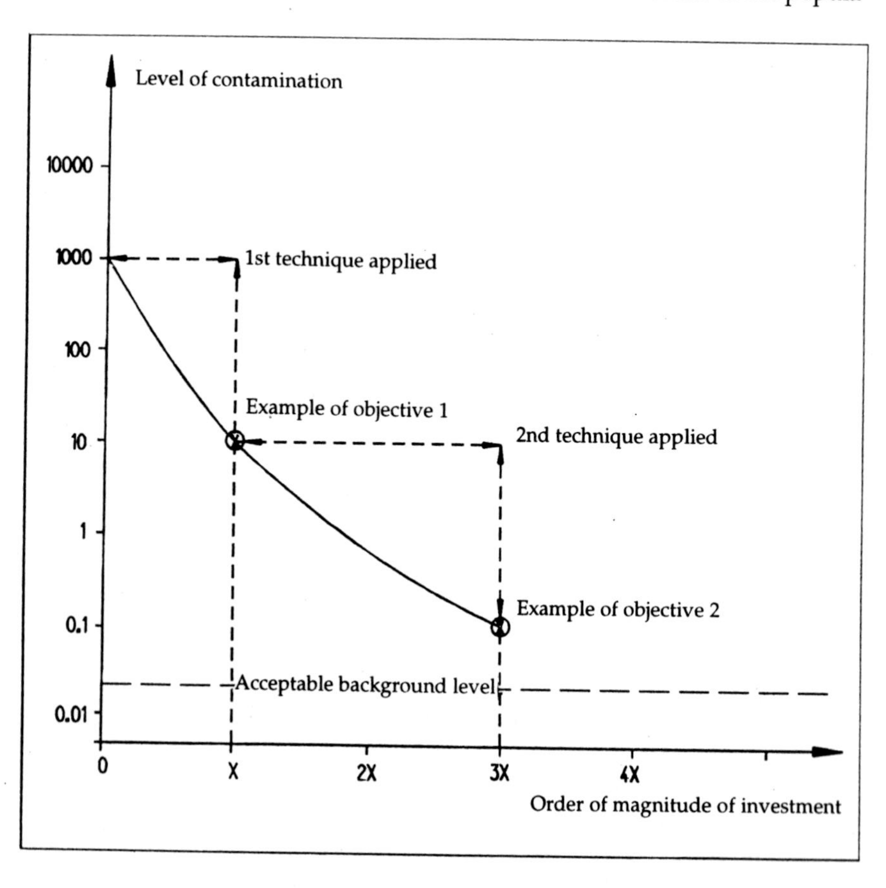

Fig. 5.1: Decreasing output versus investment.

tion, the equilibrium of local ecosystems and the technofinancial constraints.

For every case of treatment, it is essential to set objectives which, by limiting the risks encountered, can be considered 'reasonable' in the economic context relevant to the site. We saw in Chapter 4 how difficult it is to formulate and fully apply preset norms in practice. For every decision on remediation, it is necessary to define the objectives to be attained in a practical manner, by taking into account not only the norms and the risk minimisation, but also the technofinancial realities.

5.2 FACTORS INVOLVED IN SELECTION OF PROCEDURE

The choice of a plan of remediation involves a number of factors which can be divided into two distinct groups:

— technical criteria,

— economic constraints.

5.2.1 Technical Criteria

• Several aspects need to be considered at the technological level. Obviously, the proposed technique has not only to be the best suitable for the case under consideration, but also to be available in the market. Except for very special cases of trial or development of a methodology on site, one has to ensure that not only has the proposed technique been tested earlier at one or several sites (not merely in laboratory experiments), but also that it can be practically employed by agencies engaged in the treatment of remediation in the country or region in which the site to be treated is located.

Fortunately compilations giving basic information on a number of existing techniques of remediation are available today which enable a choice of the technique appropriate for the case under consideration. The procedure for utilisation of this information consists of entering the compilation in the computer, using appropriate code words, defining the main characteristics of the pollution needing treatment (for example, the type of pollutant and the type of affected component of environment, viz. water, soil etc.) and proceeding to the acceptable valid solutions in conformity with the basic category. Apart from the techniques, these databases also identify the remediation agencies and give examples of concrete cases in which applications have been successful. The two best known examples of such databased compilations are VISITT of the USA and SEDTEC of Canada. In France, a compilation of this type was recently developed by CNRSSP (Le Hecho and Marseille, 1997) which has significantly helped technological development. These tools are now being regularly used in the implementation of remediation operations.

In many cases, and particularly for biotechnologies, the feasibility of the proposed procedure will include certain preliminary and specific tests, carried out on the terrain or in the laboratory. In fact, limitation to just theoretical reflections on the method or prosedure proposed would be inappropriate and as such, may prove non-applicable in practice. For example, the indigenous bacterial flora expected to degrade the contaminant might actually be non-adaptable to this type of degradation, or the variations in permeability of some parts of the soil may obstruct employment of the proposed treatment at the scale of the entire site.

• First of all, the proposed choice should take into account the type of pollution and substances to be removed or degraded. The behaviour and characteristics of each contaminant present will obviously determine the methods selected. The characteristics of some 'standard' organic contaminants are presented in Table 5.1, revealing the differences in their behaviour that have to be taken into account when selecting the technique. Thus we shall not propose a method involving ventilation of gases from the soil for a non-volatile contaminant, nor employment of a process of biological degradation for a non-biodegradable product.

Table 5.1: Characteristics of some contaminants (after Lagny, 1991)

	Specific density	Solubility in water (in mg/l at 20 or 25°C)	Volatility (Henry's const in Pa·m^3/mole)
Eicosane	< 1	0.002	25
Hexachlorobenzene	1.60	0.11	0.005
Methyl ethyl ketone	0.81	353	2.4
Vinyl chloride	0.91	1.1	2800
Benzene	0.88	1780	555
Tetrachloroethylene	1.61	150	1500
Ethylene chloride	1.24	9200	0.97
Carbon tetrachloride	1.59	1160	2300

(Values needing special attention have been underlined)

• The technological choice should also take into account the volume to be treated and, in particular, the concentration of the contaminant in the affected medium. Each technique has a maximum efficiency for contaminant concentrations of a given order; certain methods which are more effective for higher concentrations become inoperative when the concentration is lower though still harmful. For example, the technique of pumping is highly suitable for removing an individual organic liquid contaminating substance at near 100% concentrations, but it is not possible to pump out an organic liquid 'impregnated' in the soil pores, at concentrations of several grams or several tens of grams per kilogram of soil.

• The surface area to be treated may also constitute a determinative factor for employing a particular technique versus another. Theoretically a large site of 1000 m^2 and another site of 200 hectares could both be treated by the same technique; but even when the pollution in the two cases is identical, the approach to the problem would need to be totally different in each case. Besides the surface area to be treated, the condition of the site is also an important factor; thus the access facilities and obstructions in them can play an important role in the choice of the technique of remediation. In uneven terrains or highly urbanised zones, it would be foolhardy, for example, to implement a treatment requiring a large clear plane surface for stocking mud, or to install a complex device with multiple units.

• The envisaged solution will also depend on the type of medium to be treated (soil, groundwater, ...) and its characteristics (clayey impermeable soil, very permeable alluvial, gravel...). One difficulty very often encountered is to find one or several techniques suitable for different pedological media, sometimes with contradictory characteristics and constraints. Thus every technique requiring pumping or injection of liquid or gas directly into the soil is based on the concepts of permeability and porosity. Fig. 5.2 shows the distribution of sediments as a function of their granular size (from fine to coarse) and their permeability to air; it can be seen that the coarser the grain size, the higher the permeability to air, i.e., the capacity of air to circulate in the pores of the soil. If this permeability falls below a certain threshold, it will not be possible, for example, to extract by suction contaminants present in a gaseous state in the soil. Opinions on a limiting permeability for adoption of this technique are divided in practice; some authors put the limit at 10^{-7} or 10^{-8} m/s, somewhere between clayey sand and silt, while others are more conservative with a limit between 10^{-3} and 10^{-4} m/s.

• In the course of choosing the technologies of remediation to be proposed, the specific nature of the activities on the site also plays a role; the technique used and its modalities of application will differ widely depending on whether it concerns an abandoned industrial wasteland on which the operator of remediation is totally free to organise his works, or an active factory for which the constraints of production and movement of personnel have to be honoured. Likewise, the future development of the site will exert a strong influence on the remediation operations; thus building projects which require extensive excavation work will give preference to remediation techniques involving physical removal of the contaminated material as against in-situ remediation. Lastly, along the same line of thought, the prospect of delay of future activity and/or development of the site will be an important factor governing choice.

• Another significant factor which will guide the implementation of the remediation plan concerns the administrative exigencies or those of local groups (responsible for the clean-up task). This may be interpreted not

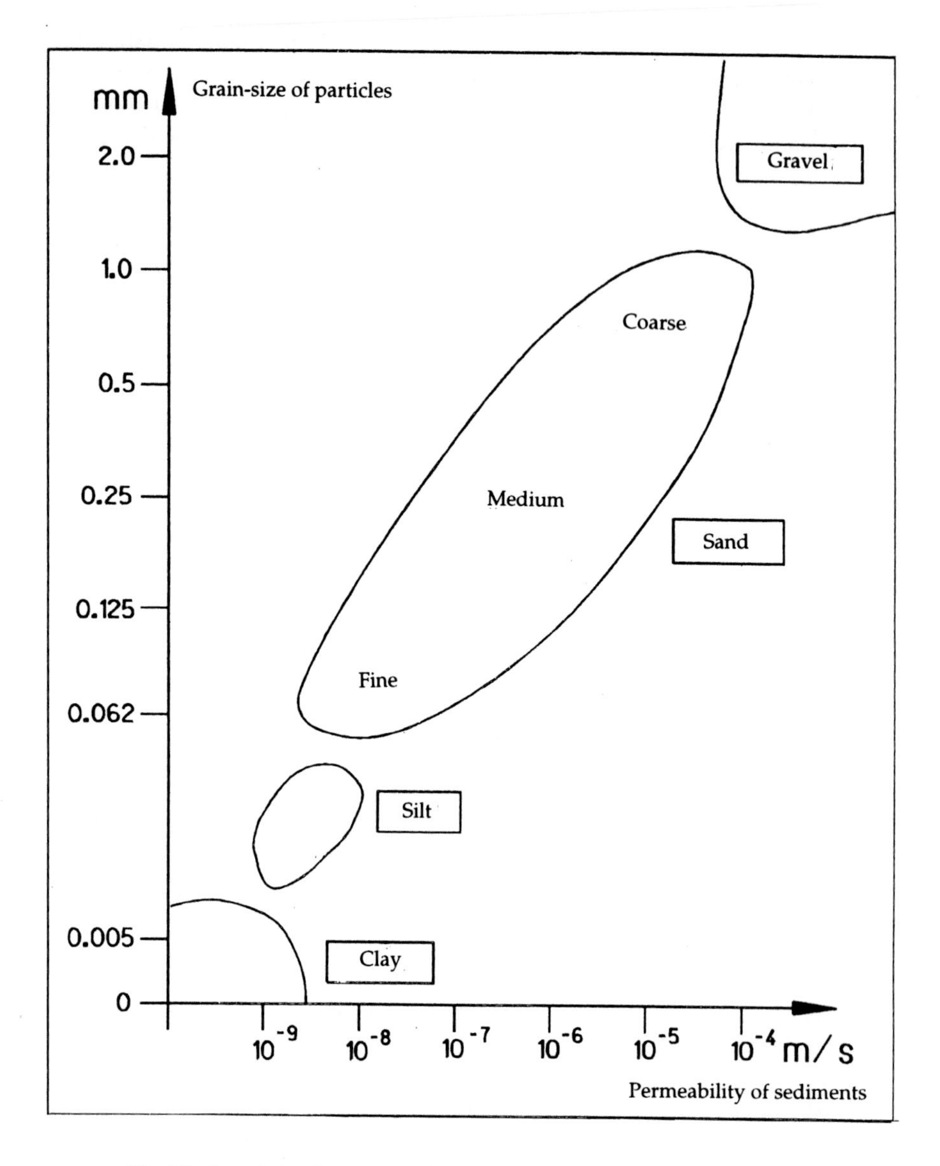

Fig. 5.2: Correlation between grain size of a soil and its permeability to air.

only in terms of the goals to be attained, for example concentration of substances in the low-lying water or soil, but also in terms of discharge at the end of the process (for example, concentration of gas in the atmosphere, quality of effluents discharged to the bodies of residual waters, or in the neighbouring water course...). Ordinarily, such exigencies orient technological choices and their modalities of application very precisely.

Concomitant with the regulations applied by the administration, the legal implications of the proposed operations also need to be estimated and taken into account, so as not to provoke conflicting situations (or aggravate one that already exists) and not to place the proprietor of the site in a legally delicate position.

5.2.2 Economic Approach

The choice of technical solutions has necessarily to be compatible with the economic aspects of the project as a whole. This entails estimating the financial cost to the proprietor of the site or the person responsible for restoring it. It would require inputs from the several operators involved specifically in the different steps of the remediation plan. An approach of this type should take into account not only the intrinsic cost of the operations themselves, but also the costs of follow-up actions, supervision of remediation and drawing up a final balance sheet to be presented to the various partners and parties concerned.

Quite often an estimate of cost drawn as above still does not represent the totality of the project because it is rarely possible before starting operations to estimate precisely the output or net-result that will be obtained from the remediation efforts and hence the time needed for realising the objectives of rehabilitation fixed initially. So, one tries to forecast the cost of remediation per unit time (per month or per quarter...), and also, per unit volume or weight of the material treated (discharge at X/ton or $/m^3$ treated per day to arrive at the daily cost Y...).

It must also be kept in mind that the cost of remediation increases exponentially with the quality of restoration desired; in other words, for a higher quality desired, the cost shall be much much higher since removal of the 'last micrograms' of contaminant is the most expensive.

To summarise, the study of feasibility and drawing up a proposal of the plan of remediation represents a complex process, involving a number of factors, often contradictory; it consists of the following:

— choosing the technical solution best adapted to removing or neutralising the effects of the observed pollution;

— planning based on as exhaustive a diagnosis and as accurate analysis of risks as possible;

— combining (in most cases) several techniques for taking into account the variations in conditions initially assumed that may be observed at various stages (substances, concentrations, media...);

— and last but not the least, making all the parties concerned aware that the operations of remediation always imply high costs and very often have to be applied over relatively long periods.

5.3 MODES OF TREATMENT APPLICATION

Irrespective of the technology used, three modes of application of remediation operations can be identified. Each has its strong points and limitations, and sometimes matches specifically one situation rather than another. But generally speaking, all can be implemented irrespective of the case under consideration.

These 'modes of treatment application' are as follows:

— application of treatment away from site,

— application of treatment in situ,

— application of treatment on or near site

Application away from site

This involves removing quantities from the natural medium by excavation or pumping the material to be remediated (earth, wastes, water...), and transporting it from the site to a centre specialised in the chosen technique where it will be treated and eventually probably brought back to the site for relaying, as in the case of soil.

This mode of operation has these advantages: the entire quantity of material to be treated is removed (one can, in the course of the operation, separate the contaminated part from the remaining 'clean' part); the treatment does not have to be done on site which may cause problems in the case of difficult sites or one on which some activity is in progress and lastly, the work performed at specialised centres is generally much more efficient.

On the negative side, in this mode of application, one is penalised by the cost of transport over sometimes long distances (several hundred km between the site and the treatment centre) and sometimes taking special precautions in conformity with the norms in force regarding the transport of hazardous material. In addition, treatment 'outside the site' increases the risk of dispersion of pollution, both during removal (for example, while excavating contaminated earth) and during transport. Finally, the material to be treated may have to meet certain specifications in order to be accepted by the chosen centre of treatment, viz. concentrations having to be within specified limits and certain harmful substances having to be absent. In this mode it might be desirable to avoid mixing and to classify the material to be remediated in terms of its composition. Thus on a site contaminated by heavy metals, one could separate the earth from which mercury has to be removed and that from which lead-zinc contaminants have to be removed, since the treatments differ, and thereby avoid a 'mercury' treatment when the two are mixed together—a treatment which would be much more onerous.

Application 'in situ'

For treatment 'in situ', the task is accomplished directly in the contaminated natural medium, without having first to separate the contaminated part. The corrective action is applied on the spot by installing the remediation system on the site and making it operate directly on the medium to be treated. This system comprises two types of equipment: one is mobile and is installed on the site above ground and is thus reusable from one site to another; the other, adapted to each particular case, consists of devices installed within the medium to be treated (i.e. the devices for injection, drilling...).

This modus operandi eliminates the operations of water pumping or earth excavation; it also precludes the transport of contaminated material.

In the 'in-situ' mode, the soil, phreatic surface, earth and groundwater can all be treated simultaneously which, undeniably, is a big advantage.

Easy to implement, this mode nonetheless involves some difficulty in control and in following the progress of the operation. In fact, when dealing directly with the natural medium, several parameters can interfere, which by definition are difficult to decipher, measure and control, imposing thereby a penalty on management of the operation. Also, it is usually difficult to estimate the efficacy of the treatment applied (the exact volume treated and the degree of treatment achieved). It is likewise not easy to discern—and take into account—the variations in level of pollution within the zones to be treated. In other words, some zones may be treated for a period far longer than necessary because their level of pollution was much lower than that estimated; on the other hand, other zones for which the level of contamination would have required treatment over a much longer period, may end up receiving inadequate treatment.

Application on or near site

The mode of operation lies in-between the two discussed above. It thus incorporates both their advantages and shortcomings.

This mode entails removing the material to be remediated from the natural medium (excavation of solid material, water pumping)—as done in the 'away-from-site' mode—but administering treatment on or near the site using a mobile remediation system that can be transported and reused elsewhere. Unlike the 'in-situ' operation of remediation, in this mode remediation is not carried out in the medium per se but rather in equipment installed at the site, which may be merely a small-scale adaptation of a large fixed treatment plant. For some techniques, it may be a system so modified that it is mobile and can be used on a small scale (several tons per day, for example).

Compared with the 'away-from-site' mode, the cost of transport is saved in this case and also the risk of contaminating the region outside the site is eliminated. Furthermore, the nuisances associated with transport, such as noise, dust, road obstacles, are precluded. Vis-à-vis the 'in-situ' mode, one is freed of control and other difficulties faced due to the natural medium; one is also able to get a good idea of the volume and tonnage involved and treated.

However, the operations of removal of material (in particular, excavation of a huge quantity of material, typically several tens of thousands of tons on a large site) constitute a difficult task in terms of management of the operation as well as financial requirements. Lastly, the cost of transport of mobile equipment from one site to another may not be negligible.

These modes of implementation are not mutually exclusive and in the course of study of the feasibility of a remediation plan, one may end up proposing the use of one or several remediation techniques, corresponding to the different modes of operation. For example, on a site contaminated by a 'cocktail' of organic substances, one may propose excavation of the most contaminated earth, sending one part to the incineration plant, treating the other part on site by the thermal method and, finally, complete the remediation by an 'in-situ' treatment by a biological process or by extraction of volatile gases.

As for costs, the incidence of the mode of application plays a definite role.

In fact, without going into the details of each technique (see Chapter 6), one can estimate that the 'away-from-site' treatment is always the most expensive amongst the three modes, partly because of the cost of transport, sometimes very large, which is absent in the other modes, but more so because recourse to large specialised treatment centres such as incineration plants, hazardous waste dumping sites etc. may itself imply a very high cost.

In contrast, the in-situ techniques are, in almost all cases, less expensive. Savings occur mainly in the costs of storage, handling and transport of materials to be treated, as also in the costs of maintenance and restoration of the site during and after the operation.

Paradoxically, the methods whose implementation is the easiest are the most expensive; contrarily, the methods whose implementation and management are the most difficult, cost less. As a matter of fact, for an away-from-site operation, one would essentially need a 'public works' type yard consisting of excavation, handling and transport to a specialised centre for treatment. On the contrary, for an 'in-situ' operation, the foreman (or manager of remediation) has to manage a complex process involving physicochemical and sometimes even biological phenomena, occurring directly in the natural medium, which are not readily accessible and observable.

present in the medium. Thus any pollutant present in the soil and contaminating the lower layers will continue to spread and 'feed' the groundwater.

If the contaminant is highly soluble in water, its dispersion through the groundwater will be rapid and the risk of contamination spreading over considerable distances is thereby enhanced until all the accumulated contaminant is exhausted.

If, on the other hand, the solubility of the contaminant is low, the substance will continue to act at the level of the groundwater table for a long period, several years, or even several tens of years. Let it be remembered that numerous organic substances of low solubility are rather toxic and dangerous even in very small concentrations (a few ppb for example—μg/l).

The efficacy of a remediation process depends primarily on the effectiveness of pumping, which per se is directly dependent on the hydrogeological and hydrodynamic properties of the medium as well as the behaviour and dispersal pattern of the contaminant. The basic problem will generally be: How to chalk out a plan of remediation by pumping to accomplish the results in a 'reasonable' length of time and within the budget provided? Experience has shown that at the time of execution, the operators have been confronted with situations not foreseen during the stage of design and feasibility study of the operation, either due to inadequate knowledge of the parameters of the medium or due to significant changes in the medium induced by the operation itself (Keely, 1989).

Figure 5.3 illustrates this phenomenon of variation of certain parameters during the course of an operation, with significant changes in the efficacy of remediation. The example pertains to the flow velocity of groundwater in the aquifer. Thus water arriving directly in the pumping well is displaced faster than that at the well's source; so some parts of the contaminated zone will be remediated quickly, others slowly. At the same time, non-contaminated parts of the aquifer, those bordering the zone of dispersion of the contaminant, are under risk of becoming contaminated by suction under the pumping action if the extraction well is located too close to the limit of the contamination. If, on the contrary, the pumping operation is correctly planned, a hydraulic embankment will be formed and the pumping will generate a cone of depression around the well, thereby precluding movement of the contaminant downwards.

The changes occurring in the flow pattern during the operation are usually too complex to be predicted or controlled, even if the designer of the project has the facility of modelling of hydrodynamic situations before, during and after the operation by means of mathematical and informatic simulations.

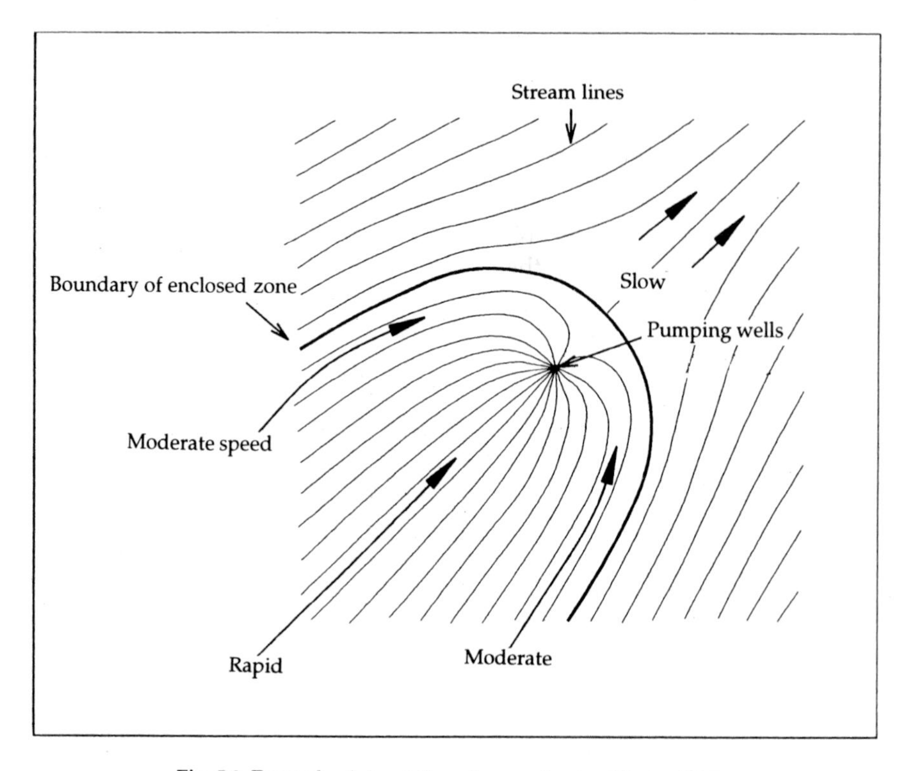

Fig. 5.3: Example of circulation of groundwater (Keely, 1989).

Resolution of a contamination problem in a potable water supply tank in Strasbourg illustrates a concrete application of pumping as a remediation technique (Lapierre et al., 1992). In this example an industrial contamination of drinking water by tetrachloroethylene originated at more than 1 km downstream (Fig. 5.4). First a diagnosis carried out by means of a series of intermediary reconnaissance piezometric wells (denoted by *P* in the Figure) helped locate the zone of entry of the contaminant and enabled accurate identification of the conditions of its transfer and flow into the groundwater. Subsequent modelling of the hydrodynamic system and various simulations of pumping made it possible to define the parameters needed for an effective pumping plan (position of well, flow rate, duration of pumping...). The fall in concentration of the contaminant in the water tank (Table 5.3) during the first months of the pumping operation of remediation revealed the efficacy of the system employed. It may be noted that the contaminated water, once pumped, was directed towards the network of urban drains and a sewage treatment plant downstream.

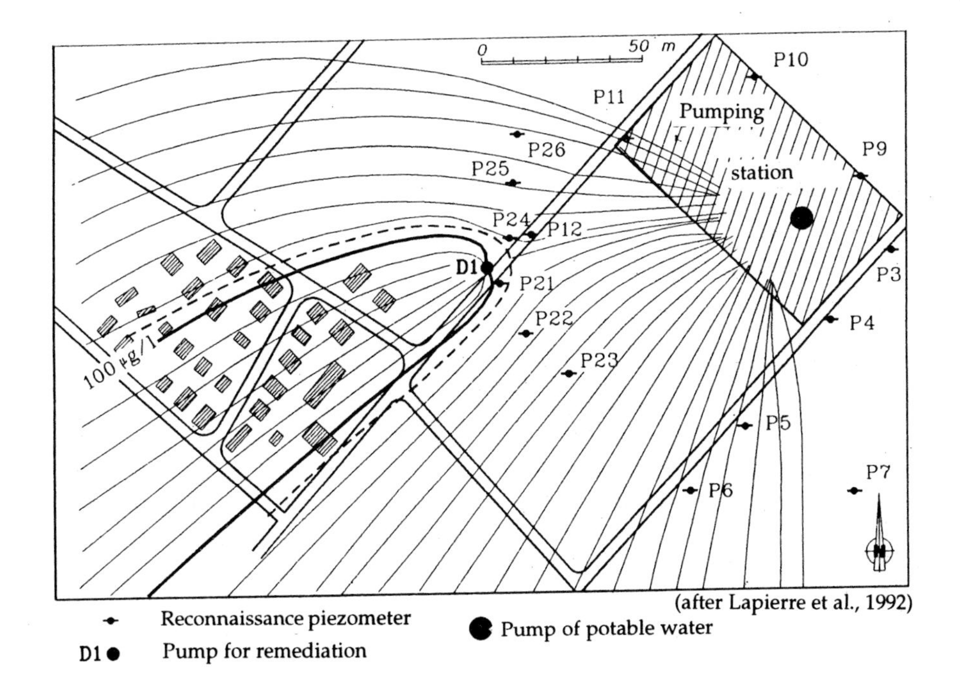

Fig. 5.4: Example of remediation of groundwater by pumping (Lapierre et al., 1992).

Table 5.3: Concentrations of tetrachloroethylene in water-supply tank and remediation well during pumping for remediation (Lapierre et al., 1992)

Date	Remediation well, average concentration in µg/l	Storage tank of potable water, average concentration in µg/l
September, 1991	–	12.7
November, 1991	142	10.5
January, 1992	111	5.1
March, 1992	70	2.3
May, 1992	51	1.9
July, 1992	20	1.3
September, 1992	18	1.5
November, 1992	21	2.1

In this particular case it may well be asked how long would pumping be necessary? As a matter of fact, and provided that the source of the pollution was identified and its propagation in the medium arrested, would it still be obligatory to pump indefinitely or, contrarily, once the concentration at the potable water-supply station has normalised, could the operation be terminated? If pumping is stopped too early, one runs the risk of the concentration starting to rise again, the stock of the pollutant contained in the aquifer not having been totally extracted. If, on the other hand, active pumping is continued, this might amount to carrying on an operation perhaps not necessary and thus a sheer waste of finances. A programme of monitoring of the quality of the groundwater would help to ensure management of the follow-up operation and resolve the dilemma; naturally this should be part of the plan and agreed to by the financing parties.

In conclusion, two points must be underscored as they are paramount in the success of an operation of remediation of groundwater by pumping:

— it is imperative to have a detailed knowledge of the hydrogeological and hydrodynamic system before designing and initiating the operation;

— it is necessary to preplan and implement follow-up procedures and those of control through a set of parameters to enable management during its progress and termination, when indicated, of the operation.

Pumping-skimming

This particular technique of remediation is applicable to organic contaminants in a liquid phase floating on the phreatic surface. As in the preceding case, a sound knowledge of hydrogeological parameters of the medium is indispensable to ensure proper dimensioning of the pumping system and thereby maximum effectiveness. A sound knowledge of the behaviour of the organic phases present in the soil is also required—ordinarily this pertains to hydrocarbons, oils and related substances; also the permeability and porosity of the soil are determining factors for the success of the operation.

When applied under favourable conditions of the medium, this method can be very effective, especially if implemented soon after a pollution occurs (an accidental spillage for example) because it helps recover in a short time the contaminants in a phase which is free, pure and has a high concentration of contaminating substances. Further, the recovery phase can be reutilised during the industrial process if it is not too degraded or soiled, after separation and filtration of the water eventually present. In case such reuse is not possible the substance will be routed to a traditional treatment unit, such as an incinerator for example.

Implementation of the technique is based on the following principles:

— When an organic phase is present in the soil, floating on the surface of the water table, several tens of cm to several metres thick, its direct pumping can be undertaken through one or several drill-holes.

— When the thickness in the drill-holes decreases to a few cm, direct pumping becomes difficult because in this case an emulsion of the organic phase and groundwater shall get extracted, which needs to be separated and treated. To offset this drawback, two pumps are installed, either in the same well or in two contiguous drillings. The first pump is installed in the groundwater itself (Fig. 5.5); water is pumped at a certain rate, calculated in terms of the parameters of the system and discharged (with or without treatment, depending on its specific quality). On pumping the water a conical depression forms in the water table around the pumping well; the size and extent of this cone depend directly on the parameters of the medium and characteristics of the pumping operation. The depression will restore the floating organic phase which tends to accumulate in the cone so formed, from which it can then be extracted by means of the second pump placed above the first within the organic phase; this is called 'skimming'.

Photo no. 4: Pumping/skimming well (source: ANTEA).

While the water has to be pumped continuously to maintain a constant cone of depression, skimming the floating organic phase has to be done cyclically, the cycle being long stretches at long intervals or short stints but in quick succession, depending on the speed of inflow and rate of filling up the cone. The starting/stopping of the pump for skimming should be automatic, controlled by level monitors installed in the well and preset manually by the person supervising the operation.

The skimming operation is characterised in terms of the duration of cycles, for example: skimming X minutes, filling of the cone, Y hours. To

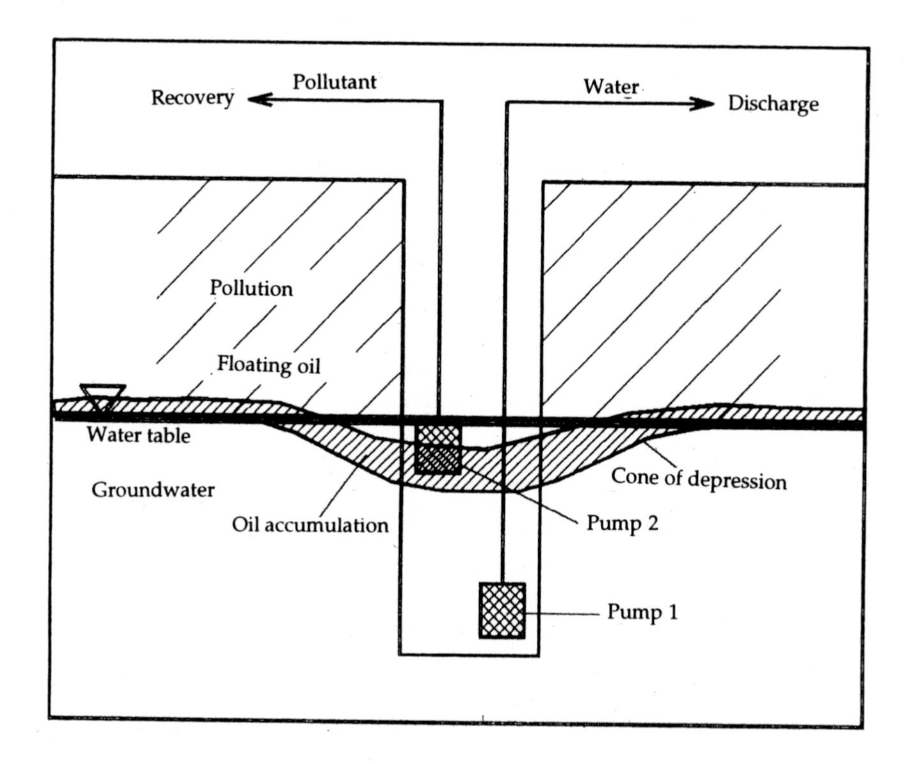

Fig. 5.5: Schematic sketch of pumping/skimming.

characterise its output one would rather use the notion of average volume of organic phase pumped per day or per month, for example: 200 litres/day, 10 cubic metres/month.

An alternative system consists of installing drains or drainage trenches at the depth of location of concentration of the contaminant to be recovered, to facilitate its collection. These drains are interconnected and lead to a pit or central well where a system of pumps is installed. A useful technique consists of installing these drains parallel to the walls staved in the soil (see next chapter), which curtails migration of the contaminant. The contaminant, by colliding against the wall, aggregates in the drainage system and is recovered through the collector wells.

Depending on the estimated quantity of the contaminant in the soil, the quantity of the substance to be recovered can be estimated by subtracting the residual phase trapped in the soil by capillary action and the phase dissolved in the groundwater, both inaccessible to the system. In this context it should be noted that the average volume extracted per day is not constant but tends to decrease as the operation progresses because concentration of the contaminant becomes more and more difficult.

Figure 5.6 illustrates this phenomenon by means of an asymptotic curve of the decrease in the rate of volume recovered, until it falls to a nearly flat rate corresponding to the maximum quantity of the substance that could be recovered in the free phase. It can be seen from this curve that it is difficult to linearly extrapolate the output of remediation on the basis of results obtained during the initial stages of the operation.

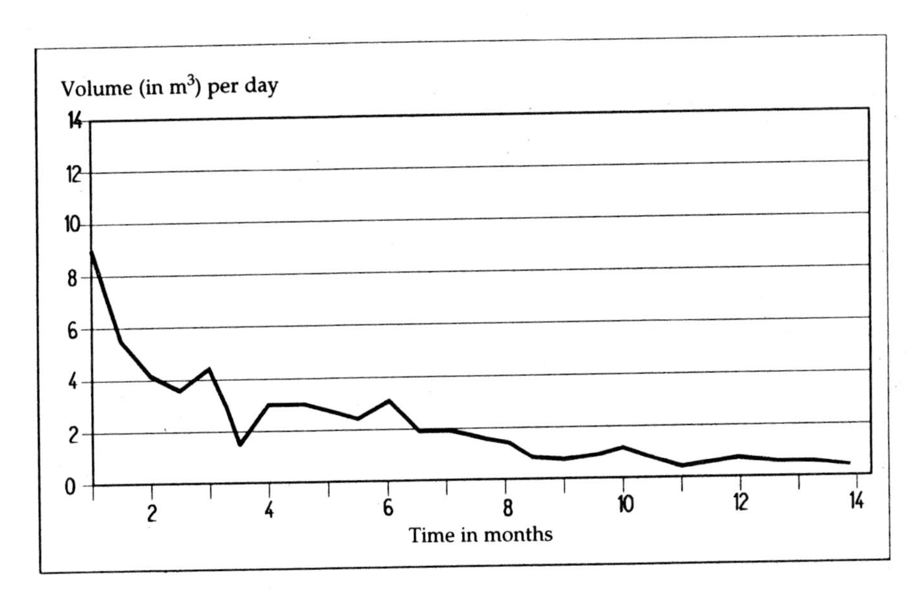

Fig. 5.6: Rate of output of pumping-skimming with progression of pumping.

This technique of skimming has an additional significant advantage: pumping of the groundwater ensures hydraulic confinement of the contaminated zone by creating a cone of depression around the drilling; this subject will be taken up in the next section.

When the thickness of the layer of floating substances is small, or one has to deal with an emulsion at the surface of the groundwater, other techniques of 'gathering' can be used, which are based on the principle of recovery of organic compounds floating on open water (rivers, sea...). It is an oleophilic (and hydrophobic) system comprised of rotating strips, drums, sponges, which absorb the oily compounds but not the water.

Several commercial forms of this system are currently available in the market, corresponding to diverse applications.

Washing and entrainment of contaminant by a liquid

To recover the residual part of a contaminant trapped in the matrix of the soil, one of the simplest methods is 'washing' the soiled earth by an appropriate liquid, thereby entraining the contaminant and restoring the cleanliness of the soil.

Photo no. 5: A well equipped with a system for recovery of floating petroleum products by oleophilic strips (source: BRGM).

Two modes of operation can be applied to this technique: either the dirty soil is excavated and washed on site, or the operation is carried out in situ by injection of a liquid that entrains the contaminant and the liquid-contaminant mixture is then pumped out. This technology is especially suitable for organic products but can also be applied to other soluble substances, such as mineral salts.

To improve recovery, two systems are possible: either detergents or surfactants are added to the water used for washing or hot water or even steam is employed.

The on-site operation may be carried out in tanks or basins. The device for separation by physical means is not easy to design. We shall see in the section titled 'Chemical Methods' (5.4.3) that the methods of extraction by solvents or by reaction with chemical solutions are in principle, quite close to the method described here.

Washing by 'physical means' consists of separating the material into fractions of different sizes and densities. In one or several of these phases there will be a concentration of contaminants (e.g., heavy metals in the finest fraction) which will require specific treatment (e.g., burial in a dump). Currently used for the treatment of ores, these techniques demonstrate the role of different unit-operations, such as sieving, flotation, filtration,.... There are several methods of treatment, which have to be adapted to the case under consideration, depending on the properties of

Photo no. 6: Treatment shed for separation by physical means (source: BRGM).

the material to be treated and the contaminants present. A standard method of treatment consists of sieving the dry material to separate out the coarsest fraction (larger than 4–5 cm). The material is then mixed with water and stirred to loosen the lumps and aggregates; the mixture thus homogenised is next wet-sieved into several fractions, separating the coarse fractions (about 1 mm to 5 cm) from the others. Here sand is separated from silt and clay by successive hydroclonic operations. The contaminants concentrate in the finest fractions which are flocculated and passed through thickening and squeezing devices before being sent for further treatment. The sand is dried before reuse. The system is implemented in a closed circuit, which collects the water, treats it, then recycles it to the flow circuit (passage through activated carbon or biofilter, for example—see discussion elsewhere). It is likewise possible to work in a confined atmosphere during the first mixing wherein the volatile compounds are detached, recovered, entrained in an airflow and treated specifically.

Flotation is considered one of the options for washing soil by physical means. Initiated by the mining industry, the technique consists of making a mixture of soil, water and surface-tension reducing chemicals and blowing air bubbles into it to trap the pollutants and while rising to the surface, forming a foam which is recovered along with the trapped pollutants. Very effective for hydrocarbons and more generally for organic pollutants, it can, in some cases, also be used for inorganic substances such as minerals.

The in-situ operation consists of injecting water directly into the medium upstream of the concentration of pollution and pumping it downstream, loaded with contaminant. The method is operative only if the soil is sufficiently permeable to allow a good circulation of fluids (e.g., sandy layers). The term 'flushing' is sometimes used for this technique.

A sketch of a device for entrainment of a contaminant by injecting/pumping hot water or steam is given in Figure 5.7. In sketch A, hot water injected upstream forms a cone of accumulation, raising the surface of the groundwater and thus washing the given zone and the surficial layer of the aquifer. Downstream, the pumped water causes a cone of depression, drawing the liquid towards the depression. Between the two appears a front of hot water moving from upstream to downstream, entraining the contaminating phase trapped in the soil. In sketch B, steam is injected directly into the groundwater, heating the water and causing formation of a hot front, also entraining the contaminant; this then moves towards the cone of depression created around the pumping well where it is recovered.

The two variations—hot water and steam—can be used together, resulting in a cumulative effect. Entrainment in steam is especially advisable when dealing with products of low solubility, denser than water which, instead of floating on the surface of the groundwater, contaminate the entire height of the aquifer by migrating under the effect of gravity (families of DNAPL).

The main difficulty in this type of treatment lies in perfect mastery of the hydraulic system thus created. In fact, injection of water upstream of a contaminated zone involves the risk of dispersion of the contaminant and its fast spread downstream or laterally with respect to the axis formed by the injection and the pumping wells, essentially outside the planned zone of treatment.

Two precautions are necessary in all cases employing this method:

— First, a sufficiently dense network of piezometers should be installed all around the zone to be treated, to ensure confinement of the contaminant within this zone.

— Second, recovery wells should be located at a relatively small distance from the injection wells, determined in terms of the characteristics of the medium and generally in the order of several metres to several tens of metres at the maximum.

For either hot water or steam, an important control parameter for the progress of the contaminant dispersal phenomenon lies in watching changes in water temperature, both in the monitoring piezometers and the recovery wells.

Although the technique of direct injection into the groundwater is difficult to implement, it nevertheless has two advantages:

— the cone of accumulation of water which is generated, aids washing part of the soil that is normally not submerged;

— the dynamic effect of the injection/pumping system ensures treatment of larger zones, compared to simple pumping.

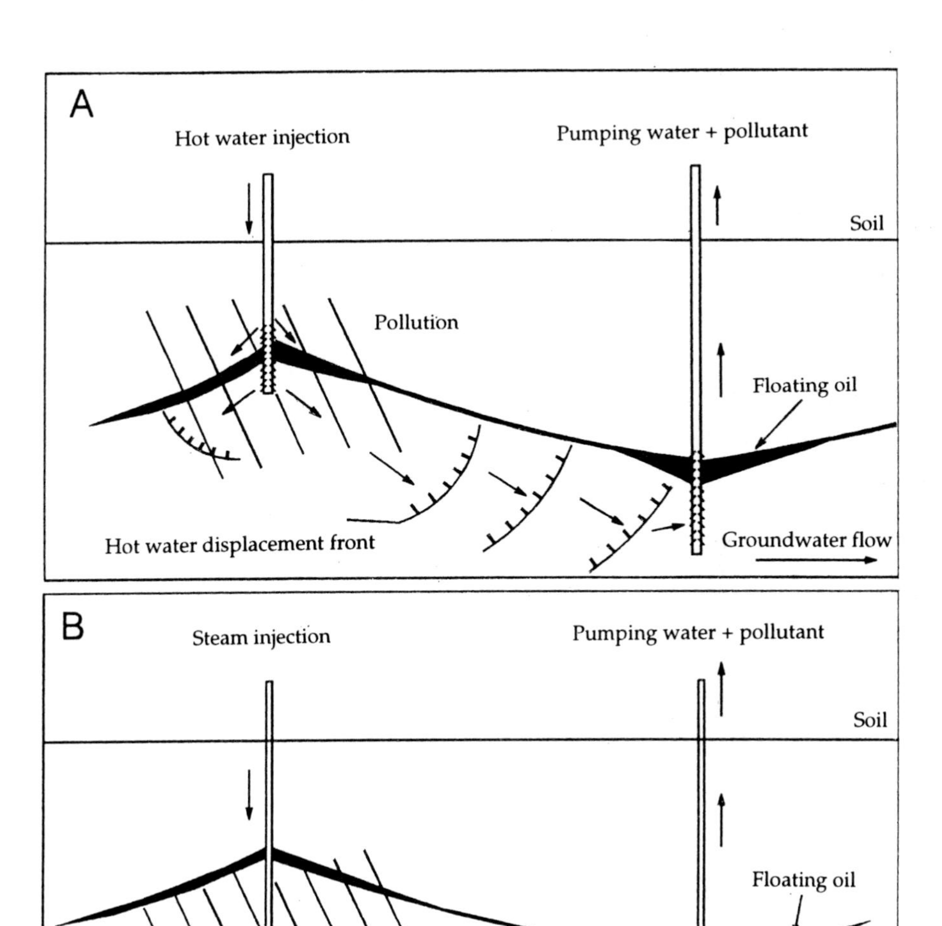

Fig. 5.7: Schematic sketch of in-situ entrainment. A—water; B—steam.

Extraction of gaseous pollutants

In-situ venting

Extraction of volatile contaminants contained in a soil—generally termed 'venting' by Anglo-Saxons—is a recently developed and successful technology used in Europe as well as North America, because of its high output and relatively low cost of implementation. The technique is particularly effective for treating cases of contamination by light hydrocarbons such as petrol in particular and volatile compounds in general.

It is a technique applied 'in situ' which can treat not only the unsaturated part of the soil (above the water table), but also the water table itself—termed 'stripping' or 'sparging'.

The principle consists of passing an airflow into the soil by means of a series of pipes equipped with strainers (Fig. 5.8), generally located on the boundary of the zone to be treated, and pumping out air loaded with contaminating gases from the contaminated zone. The airflow is generated by a system of blowers at the entrance and exhaust fans at the exit. The volatile products are thus entrained by the air circulating in the soil and extracted through suction holes. The contaminated gases are treated at the exit before releasing the exhausted air into the atmosphere.

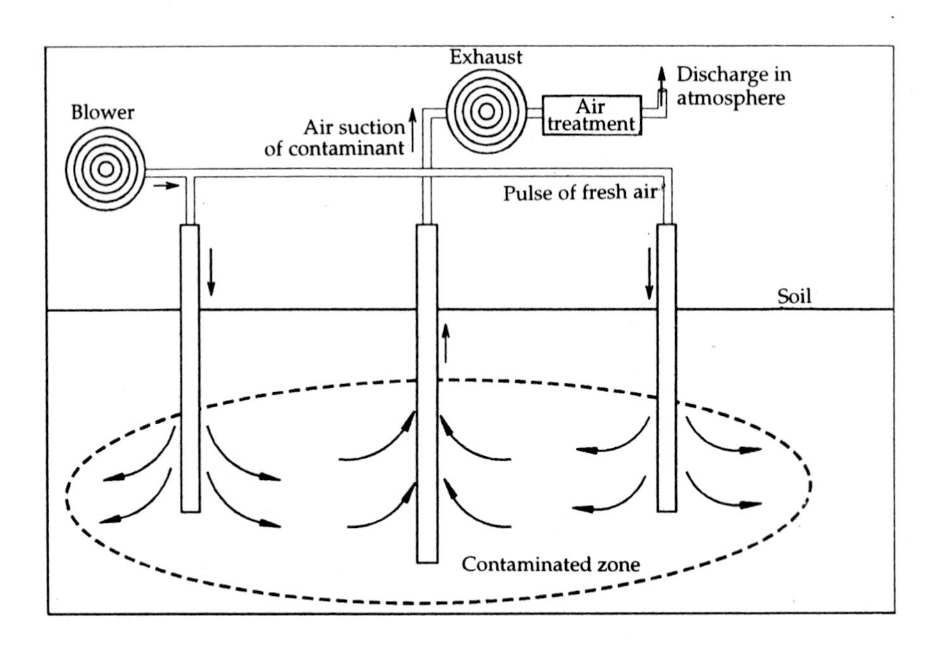

Fig. 5.8: Schematic sketch of the process venting.

Volatile organic substances present in the soil tend to evaporate and saturate the voids by their vapours; thus soil saturated with petrol may contain as much as 25,000 ppm of the product in the gaseous form. Once this phase has been extracted, the flux of pure air injected will facilitate volatilisation of the residual liquid phase. At commencement of the operation the quantity of gas extracted is large and the high concentration of gas in the soil will rapidly decrease exponentially over time (Fig. 5.9). On stopping the circulation of air, one observes an augmentation in concentration of gas in the soil, corresponding to evaporation of part of the remaining liquid contaminant. On resumption of venting, the quality of extracted gas is again large and decreases rapidly. After a second stoppage, the quantity of gas extracted is smaller than in the preceding case, and thus the cycle of extraction is repeated until the residual concentration attains a small steady value, indicating complete remediation of the medium.

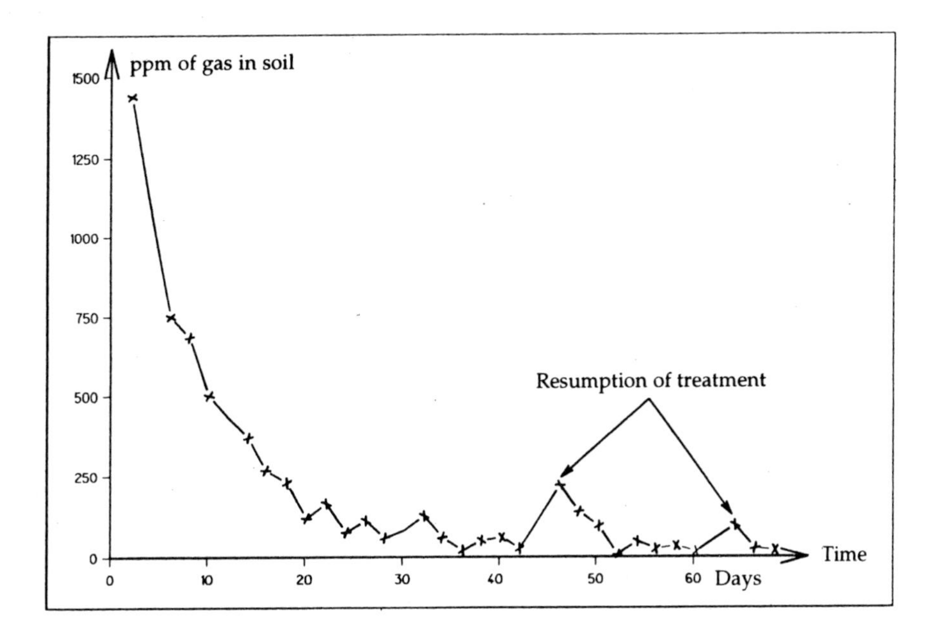

Fig. 5.9: Output of a venting operation: change in concentration over time.

This technique is applicable only when the contaminant is sufficiently volatile so that it can be recovered in gaseous form. It is generally believed that compounds with a vapour pressure higher than 1 mm mercury (at 25°C) or Henry's* constant larger than 0.01, are susceptible to extraction by venting. Table 5.4 shows the vapour pressure and Henry's constant of some standard organic compounds; it can be seen that substances such as benzene or trichloroethylene, readily volatilisable, are particularly suitable for remediation by venting. It can also be seen that the method is likewise applicable to substances such as naphthalene or phenols, which have a

*Henry's constant corresponds to the ratio of the equilibrium concentration of a substance in the gaseous phase to its concentration in the soil pore-water. Thus the larger the Henry's constant, the faster the volatilisation of the compound.

Table 5.4: Examples of Henry's constant

Compound	Vapour pressure (in mm of Hg)	Henry's constant (dimensionless)
Benzene	95.2	240
Toluene	28.1	330
Trichloroethylene	57.9	450
Carbon tetrachloride	90.0	1,300
Vinyl chloride	30.0	390,000
Phenol	0.353	0.016
Pentachlorophenol (PCP)	1.5×10^{-5}	0.13
Naphthalene	0.054	22

small Henry's constant that is nevertheless larger than 0.01. Such compounds are termed semi-volatile. Obviously, the procedure will be less effective in this case, all other factors remaining constant, and will generally have to be applied for a longer period of time.

Apart from the nature of the contaminating product, some other factors should be taken into account during a study of the feasibility of venting and efficacy of the operation, namely: the characteristics of the soil, such as its lithological composition (is it composed of sand, clay...?), its structure (how are the particles distributed, what is their size...?), humidity, permeability, porosity.... Thus venting is more directly applicable on a highly permeable soil in which the distribution of particles is uniform. Permeability with an average value higher than 10^{-5} m/s represents a minimum for application of venting, although remediation by extraction of gas has been successfully carried out for soils with a permeability even of the order of 10^{-8} m/s. Success of the method is better guaranteed in zones of gravel and sand in an alluvial plain than in clay on the summit of a chalky plateau.

The layout of the site and distribution of the pollution are also deciding factors in the choice of venting. The method is particularly recommended when the pollution extends over a large area and to a great depth or, under buildings or immobile surfaces (railway embankments for example).

One of the essential points to be considered during the planning of the project is the treatment of gases at the exit of extraction. Several plans can be envisaged but the final decision has to be taken in consultation with the administration and in accordance with the legislation in force. Direct discharge is normally not admissible except in the case of very low concentrations of the contaminating gases and very small quantities to be discharged. The treatments mainly include adsorption over activated carbon, incineration, passage over a biofilter or condensation.

Adsorption over activated carbon and incineration are currently the most frequently employed treatments; the former, suitable for all types of substances, is preferably reserved for chloro-solvents, while the second is used in particular for burning petrol vapours or aromatic substances (BTX, benzene, toluene, xylenes).

Examples of success are numerous. Thus, on the site of Verona well in Michigan (USA), which was highly contaminated by chloro-solvents and BTX, remediation by venting lasted three years (EPA, 1990). Initially the concentration of each contaminant present in the soil exceeded 1 g/kg. Between March 1987 and October 1990 venting was done for 400 days and almost 100 tons of contaminants were extracted. After just 125 days of operation, the quantum in the soil had reduced to between 1 and 12 mg/kg, depending on the substance; after 400 days the concentrations were lower than 0.1 mg/kg, which corresponded to the remediation objective stipulated by the administration.

Numerous variants of the above standard process of venting exist, consisting of modifying some aspects of traditional venting, or combining several devices involving several techniques (see the subsection 'Bioventing' under the section on Biological Methods, 5.4.5), or even treating several phases or several media simultaneously. In many cases the operators try to promote their particular system by attempting to describe the best results obtained in terms of the cases they have handled. An exhaustive review of these systems is beyond the scope of this book; we shall limit ourselves to a few variants which are of particular interest in terms of principle and application.

Variant one: Extraction by air vacuum (or SVE, soil vapour extraction)

This procedure does not involve injection of air, but only extraction of gases from the soil. A low pressure of the order of several millibars is created, which induces a strong suction of the gaseous phase present in the soil, the effect of which is felt over a radius of several metres to several tens of metres depending on the properties of the soil (nature, structure, permeability...), and which entrains, as a matter of course, the volatile contaminant present. A schematic sketch of this technique is given in Figure 5.10. Gases are treated as in the case of standard venting, before being discharged to atmosphere. The Figure shows an incineration oven coupled with an extraction pump on a mobile assembly, providing considerable flexibility of use and transfer from one site to another.

Accidents at Chavanay and Voulte (Ardèche) caused a strong petrol pollution of the soil which BRGM successfully remediated by using the technique of extraction of air vacuum (Antoine et al., 1993a, b).

At Chavanay, following the derailment of a train of tanker-wagons (December, 1990), between 250 and 300 m^3 of petrol infiltrated into the soil and contaminated the water table. In addition to employing a hydraulic trap (see next section) and pumping/skimming the phase floating on the phreatic surface, remediation of the soil was accomplished by air-vacuum operation, by means of 90 suction wells connected to a network at low pressure (250 millibars).

The concentration of gas observed in the contaminated zone reached several thousand ppm after the accident. The goal for remediation of the soil, fixed at 10 mg/kg, was attained in May 1993, i.e., after 14 months of treatment.

Photo no. 7: Equipment for on-site venting (source: BRGM-VALTECH).

This technique has its limitations and is applicable only to the soil above the water table. Also, when the groundwater is close to the surface (at a depth of less than a few metres), the depression created in the soil by exhaust induces suction of water in large quantities into the collection system, disturbing the system and lowering its efficiency.

In view of this fact, manufacturers have significantly improved the machinery of extraction. What was originally considered a handicap, now turns out to be an advantage by providing a forced simultaneous extraction of air and water; also, a separation mechanism enables collection of pol-

of the groundwater. Like other venting techniques, it takes into account all volatile and semi-volatile contaminants. It is called IEG-UVB (*Unterdruckverdampferbrunnen*), literally 'vaporisation/depression well'.

The uniqueness of the method is found in the design of the well in which treatment is carried out (Fig. 5.12). It is fitted with two double-walled strainers; their respective location along depth is determined relative to the depth of the contamination, characteristics of the sediments and the medium to be treated, say for example, the top layer of the groundwater and the depth of the aquifer. An impermeable circular plug is inserted between the two strainers, isolating the two parts of the hole. At the surface, a pipe is connected to the air exhaust. Air sucked from the surface penetrates the well through an interior pipe connected to a 'stripping chamber'. Exchange between the fresh air and the contaminated media—air of the soil and/or underground water—takes place here. Groundwater gets sucked by the gaseous flow through the bottom strainer and is pushed back into the aquifer through the upper strainer. Water rises from the bottom strainer towards the stripping chamber through an interior pipe. When the permeability of the medium is small ($< 10^{-6}$ m/s), an additional pump is installed at this location to facilitate water rise. Air circulating in the upper part of the well becomes loaded with volatile compounds during the 'stripping' process and is recovered at the surface, where it is then treated before discharge into the atmosphere.

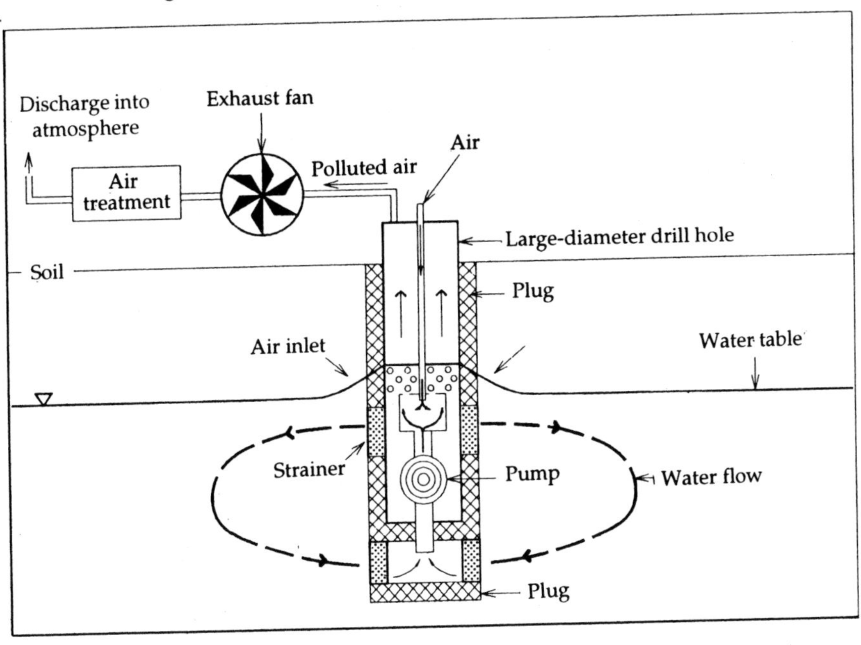

Fig. 5.12: Schematic representation of the IEG-UVB in-situ technique.

Thus two cross-convection currents are created: one by the suctioned airflow and the other by the underground water entering and exiting the hole at the levels of the two strainers. The contaminated air of the soil mixes with the fresh air entering by suction at the level of the upper strainer. It should be noted that because of this convection current, the treatment is effective within a certain radius around the well, both in the groundwater as well as above it, in the unsaturated part of the soil.

While designing the project, the zone of action and number of wells to be provided have to be determined according to the nature and extent of the pollution as well the characteristics of the terrain.

On the site of a metallurgical factory situated in Germany (Heerling et al., 1991) chloro-solvents (mainly trichloroethane and dichloromethane) contaminated the soil and the water table to a depth of ten metres around wells into which liquid residues of the factory were allowed to flow. Initially, the concentration of solvents in the water (all substances taken together) was of the order 1000 to 3000 µg/l in the contaminated zone, and several hundred µg/l at a few tens of metres downstream of this zone. After 16 months of treatment the concentration dropped to less than 10 µg/l, the goal of remediation having been fixed at 50 µg/l, and about 1300 kg of contaminating substances were extracted. Six months after completion of the operation, the concentrations in groundwater, regularly checked in the zone and downstream, were found constant at this low level.

Variant four: 'Slurping'

The combination of the two methods of extraction of the gaseous phase from the soil, one by sucking under vacuum (SVE) and the other flushing by air under pressure, is called 'slurping'. In the case of semi-volatile compounds, this technique enables improvement in results obtained separately with the two methods.

As with the two basic techniques, the permeability and the structure of the soil continue to remain the major controlling factors.

5.4.2 *Physical Methods of Trapping Pollution*

The common general objective of all trapping methods is to freeze a pollutant at the site where it begins to infiltrate the natural medium, to preclude its further spread. The strategy is to arrest the eventual migration of contamination, thereby suppressing risk to the population living around the site and the local ecosystem, given the assumption that the source of pollution is confined to the site. So, in these techniques the contaminant is not destroyed but, rather, enclosed and locked in the natural medium. By definition, these techniques are applied in situ, although some of them, stabilisation for example, may in certain cases be applied on or away from the site.

In general, the advantage of such techniques is their easy implementation; compared to chemical, thermal and biological methods, they usually require a lower degree of technical and engineering skills and are sometimes relatively less onerous in terms of volume of terrain involved.

A major question arises, however: **what is the durability of such a system after implementation? Is it not merely a temporary solution**, while removal of contamination and rehabilitation of the site are envisaged as more or less long-term solutions? At present, not much is known about the life-duration of confinement or stabilisation. Operations of this type implemented some 29–30 years ago have shown that confinement often needed periodic 'touch-ups'. As a matter of fact, total isolation of an outdoor medium is difficult to obtain and, more particularly, to maintain for a long time. Nevertheless such techniques are continuing to evolve (Géoconfine, 1993); the means and tools available today, as well as the precautions exercised during a confinement or stabilisation operation, are certainly superior to those of 20 years ago. Furthermore, the technical advances evolved for confinement of radioactive materials are now beneficial for all other families of chemical compounds also.

In France, very strict laws pertaining to the subject of norms of confinement have been recently promulgated (see Chapter 7 pertaining to legislative regulations), and in most western countries also.

Whatever the technique employed, it is evident that every type of confinement necessitates follow-up action and regular periodic checks, implying a system of supervision of the quality of the environmental medium (air, water, flora...) and regular and strict procedural measures, as only these can guarantee confidence over a long period of time.

This group of techniques comprises **three main technological alternatives**:

— confinement by encapsulation or alveolation,
— stabilisation and enervation,
— confinement by hydraulic trap.

The principle and relevant applications of each of these are briefly described below. A subsection is devoted in particular to the problems of dumping and the question of watertightness.

Confinement by encapsulation and alveolation

Such confinement consists of physically enclosing the contaminated material in an enclosure with watertight walls, cover and/or bottom. Especially developed for storing wastes or covering dumps, the methods of encapsulation and alveolation are now often applied to the treatment of contaminated earth.

Three modes of application may be mentioned here: placement in refuse dumps away from the site, 'entombing' on the site a volume of excavated contaminated earth, or lastly, covering a contaminated zone in situ and making the enclosure watertight. These techniques are applicable to all types of contaminants and are particularly useful for multisubstance or complex contaminants when the level of concentration is high. With reference to technical and financial considerations, they are more suitable for relatively limited volumes and surfaces, and for in-situ application, provided the contaminated zone is well defined and delimited beforehand.

The main objective is to freeze migration of the contaminant(s) (Fig. 5.13a); several **criteria of dispersion** must be taken into account for success of the operation.

The first criterion to be honoured for success of the confinement, is to prevent the entry of water into the zone to be enclosed. To suppress infiltration of surface water (rain, streams, ...) it should have a watertight cover. Laterally, the 'enclosure' must be closed to water coming in from upstream or from around, especially if the contaminated zone or dump is in contact with the groundwater. For this purpose, watertight walls have obligatorily to be provided. Two cases (Fig. 5.13b and c) may arise: either the zone to be confined is completely enclosed (types of walls are described later) or a wall is inserted between the zone and the hydraulic upstream to make a barrage for the redirected water.

A second criterion for this type of confinement is to suppress the risk of downward migration of contaminants. If the operation is conducted after excavation of the earth, the pit should be built in an alveolus with watertight bottom and walls (Fig. 5.13d); if, on the other hand, the operation is carried out in situ, a watertight lateral wall is provided, which is anchored in a locally existing impermeable geological layer (Fig. 5.13e). At present, installation of such walls can be accomplished up to a depth of about 10 metres, even more in certain applications. In the absence of an impermeable layer, care should be taken to embed the lateral wall to a depth greater than the base of the contaminated zone (Fig. 5.13b).

Lastly, the third criterion to be considered is control and recovery of the leachates and gaseous emanations formed in the interior of the confined zone. In an in-situ encapsulation, this operation is conducted by means of wells and vents passing through the watertight top-cover in which one can also take measurements and, if need be, recover through pumping the gases and (partially) the leachates formed. In the case of entombment, drains are installed at the bottom of the alveolus in a layer of permeable material and connected to one or several wells open to the exterior; a system of drainage (permeable layer) and vents likewise helps in recovery of gases (formation of methane in a refuse dump for example). Fig. 5.14

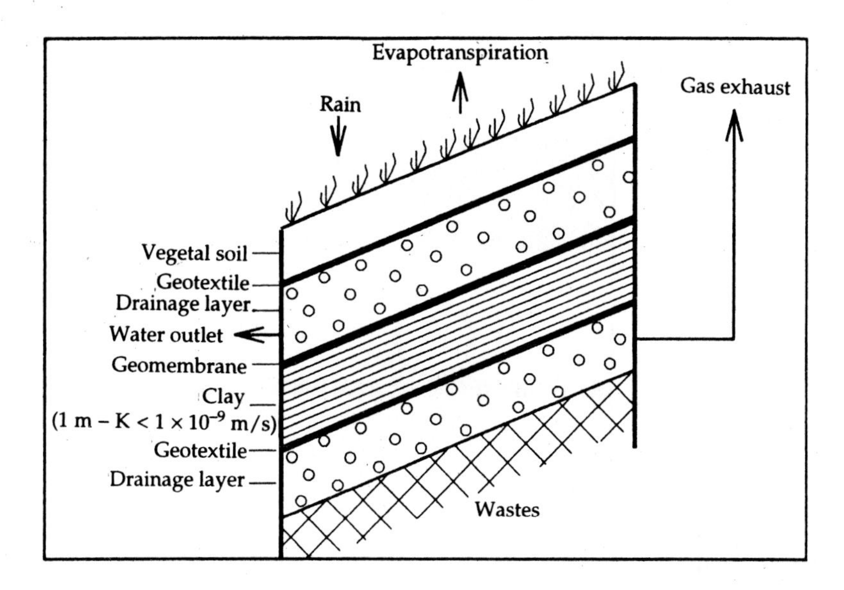

Fig. 5.15: Example of typical cover (source: V. Faucon, ANTEA).

The basis of attaining watertightness is similar for the walls or bottom (Fig. 5.16); on the impermeable natural bottom (layer of clay for example), termed the 'passive' barrier, a drainage layer is laid, followed by the geomembrane. Above it, another drainage layer, thicker than the one below the geomembrane contains the network of drains. The ensemble is protected by a geotextile, which lies in direct contact with the contaminated material.

Pitfalls to be avoided

Subject to climatic conditions and various physical constraints, the confinement must be carried out keeping in mind **three major factors, to preclude damaging its resistance and watertightness**:

— settlement,
— desiccation,
— freezing.

Phenomena of settlement can lead to fracturing of the layers assembled and thus imperil the watertightness of the dump. In fact, differential settlement due to various causes, say vehicular traffic, can, if extensive, notably diminish the effectiveness of the cover.

Right at the stage of design, the stability characteristics of the various components of the cover, bottom and the store itself, must be clearly specified and during assembly, laying and compaction carried out with maximum care. In general, settlement is quick but can be effective over several years; a temporary solution is sometimes preferred, which consists

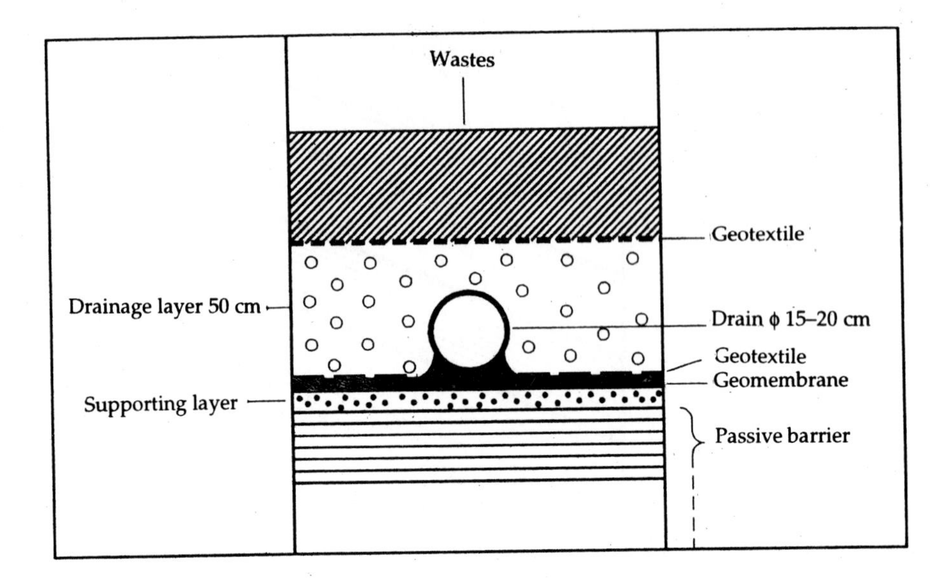

Fig. 5.16: Example of the bottom layer of confinement (source: V. Faucon, ANTEA).

of a provisional cover over the dump to protect the immediate environment and then improving the cover once the process of settlement is complete. Another solution consists of predicting the zones where settlement will be maximum and using a suitable material there. Thus, BRGM (Laffitte et al., 1985) designed and carried out encapsulation of a petrochemical refuse dump at Moselle, in which the zones sensitive to settlement (vehicular traffic) were treated by depositing fly ash three times at intervals and by adequate sloping of embankments, before the final impermeable cover was obtained.

Desiccation may occur during or after installation of a confinement; in the process the layers of impermeable sediments may crack and thus become partly permeable.

While the work is in progress it has to be ensured that a certain level of humidity is maintained in the material used, by adding water as needed.

To prevent desiccation after the installation has been completed, two alternatives are possible for minimising the process: either an impermeable layer is laid at a sufficient depth, calculated in terms of its own characteristics and local climatic conditions or, if this depth is too large for fulfilment of technical constraints, the impermeable layer may be covered with a geomembrane.

In certain regions **freezing** may tend to destabilise the impermeable plug. It has been observed that a few freezing/defrosting cycles suffice to diminish the coefficient of permeability by a factor of 100, thereby causing a strong perturbation in the watertightness of the cover. As in the case of desiccation, precautions must be taken while the process of confinement is in progress, to avoid the effects of freezing. One may, for example, install the impermeable layer and cover it with protective materials (natural soil, drainage layer or a neutral layer) up to a sufficient height.

Stabilisation and enervation

In terms of their objectives and applications, methods of stabilisation and solidification are very similar techniques. The first consists of transforming a relatively soluble substance into an insoluble compound through a chemical reaction or by adsorbing it over a neutral matrix. The second consists of mixing the contaminating substance with various additives to obtain a solid composite material, non-reactive and with low permeability. Whatever the method, the contaminants are not destroyed but their potential impact on the environment is considerably diminished because they are immobilised and physically (or chemically) associated with an immobilising matrix. These techniques represent 25% of the remediation operations carried out in the USA (EPA, 1993b).

The number of currently available procedures of solidification and stabilisation is quite large, and very often the operators have developed (and patented) their own particular techniques.

The most usual techniques consist of mixing cement (or another cementing material) and water with the contaminated products (earth, waste...) to obtain a hard material, remaining totally insoluble and stable over time, in which the contaminant is trapped.

A second procedure, also very useful, consists of enervating (rendering inert) the contaminated material by mixing it with lime or ash; stabilisation is produced by the liaison of the phase to be rendered inert with the calcium of lime or the non-silicate part of ash. This technique is frequently employed, for example, for relatively viscous hydrocarbon phases (enervating the muds of lagoons).

Use of asphalt is another option in stabilisation; the finely powdered material is mixed at high temperature with the asphalt and becomes enervated when cooled. The volatile compounds are then recovered and treated separately.

Finally, mention may be made of using various adsorbing materials, neutral and non-biodegradable, to eliminate contaminants from the liquid phase; these are primarily clay, zeolites, activated carbon and even sodium and calcium silicates. For example, thin sheets of clay, dipped in ammonium ions, are effective for retaining organic compounds trapped through binding with the negative electric charges which line their surfaces.

This list is not exhaustive. Irrespective of the actual technique employed, either the additives are directly mixed with the contaminated medium excavated after taking it on the surface, or applied by in-situ injection. In the latter case the difficulty lies in obtaining a mixture that is homogeneous and adequately packed with the stabilising substance injected in the contaminated soil. A procedure, presently operational at a few sites, was perfected by the American firm S.M.W. Seiko (Anon., *The Hazardous Waste Consultant*, Sept/Oct. 1992); it consists of injecting a mixture with cement base into the contaminated zone through a series of auger drillings. The stabilising mixture is introduced by the drill tube-auger and thus the mixture penetrates the soil. When the tool is retrieved, a cemented column (Fig. 5.17) is obtained. The drillings are close-set and cover the entire zone to be treated. This procedure can also be used for constructing the walls of confinement around a contamination. It facilitates treatment of the soil up to a depth of about 30 metres, with an output of the order 75 m^3 per day.

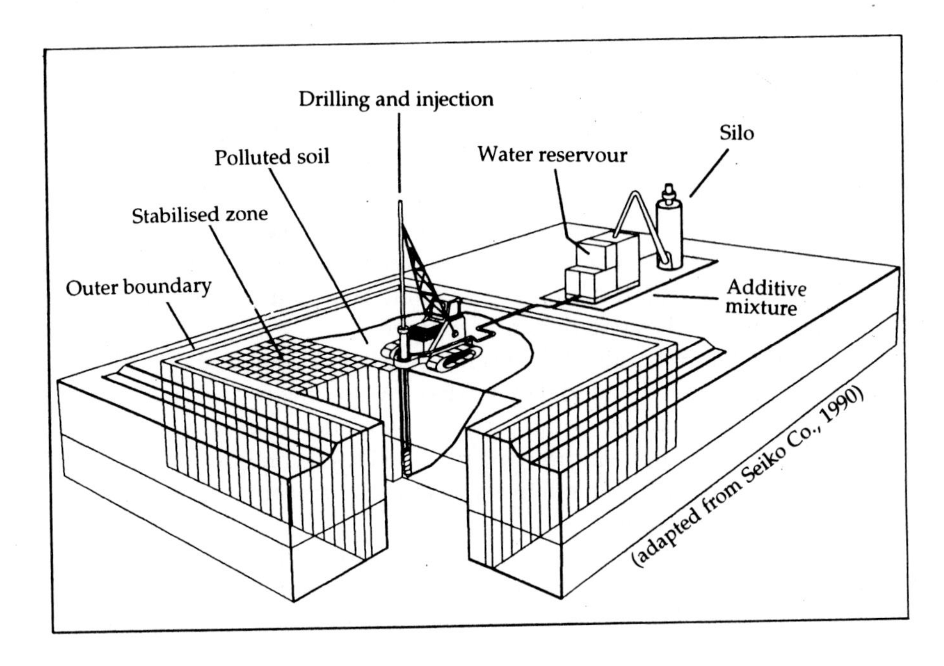

Fig. 5.17: Schematic sketch of in-situ enervation procedure of S.M.W. Seiko firm.

The techniques of solidification and stabilisation are applicable to different types of contaminants.

Solidification by cementing is suitable for earth contaminated by heavy metals. Numerous tests, conducted particularly in the USA, have shown that the quantity of metal measured in leachates of samples solidified in this manner, is very small compared to that obtained in leachates from untreated soils; for elements such as zinc, arsenic, lead, cadmium or copper, one finds several tens or hundreds of mg/l in leachates from untreated soils against less than 0.1 mg/l in leachates from solidified samples—a reduction of more than 95%.

Stabilisation by lime or clay can be effective for hydrocarbon compounds or chloro-solvents; the technique can also be applied to PCB, creosotes or pesticides. For these substances also the results show a large reduction in leached quantities after treatment. In the case of solvents or hydrocarbons, there is a risk of evaporation of a part of the contaminant during the operation; there is also the risk of polluting emissions and special precautions have to be taken to preclude their direct release into the atmosphere.

Solidification by in-situ vitrification is a somewhat different concept; it will be described in the section devoted to thermal methods (5.4.4).

Also, mention needs to be made of the recent legal obligation of stabilising the specified industrial wastes or refractory wastes before disposing them off as discharge of class 1. This is illustrated by Bouchelaghem et al. (1995) in a case of the method of stabilising wastes containing heavy metals involving a high degree of fixation of metals in the stabilised mass at the cost of relatively small increase in the final volume—essential criterion for the operation. As a matter of fact, the figures show that the soluble fraction of wastes changes from 20–70% to 2.1–5.9% after stabilisation and no contaminating metals (Zn, Pb, Cd, Cr and Hg) are detected; after the treatment the volume of a ton of residue increases from 0.97 to 1.40 m^3.

To conclude this discussion of methods of solidification/stabilisation, we give a résumé of their specific advantages and limitations.

Their essential advantage lies in suppression (or considerable reduction) of the risk of migration of pollutants to the natural environment—they become almost immobile. In addition, these techniques do not necessitate follow-up action nor attention over a long time nor additional treatments (management of infiltrations or emissions for example). Finally, the additives utilised are generally those currently in use, easy to produce or obtain, and thus relatively less onerous.

On the other hand, the methods of solidification/stabilisation have certain disadvantages or pose specific difficulties. Addition of enervating products causes an increase in the volume of the material, sometimes large and particularly penalising in in-situ applications.

When the technology is used for organic compounds, it poses additionally the risk of volatilisation of part of the pollutants or, for in-situ operations, even spread of the contamination to the area outside the zone of contamination at the time of injection.

Finally, it may be recalled that if a system of injections is employed, one encounters difficulty in achieving effective penetration throughout the entire zone to be treated as well as obtaining a homogenised mixture of additives and contaminated soil.

Hydraulic trap

In the techniques of confinement—the term hydraulic confinement is also used—a hydraulic trap may be considered a technique that closely resembles those of pumping and skimming (described in the earlier section) and complements their effects.

Frequently used, its installation and execution are relatively simple and hence not very expensive (see chapter dealing with costs). The trap presupposes, however, a sound knowledge of the hydrological and hydrodynamic properties of the medium and meticulous management of the operations of pumping employed (flow rate, time...). Furthermore, it is obligatory to obtain authorisation to pump water into the natural medium or to discharge water towards a network of drains/sewers.

Principle of functioning and follow-up (Fig. 5.18)

— Following contamination of the subsoil and water table, the contaminating substances tend to migrate downstream along with groundwater, depending on their characteristics, and may 'exit' or go out from the site of contamination. Figure 5.18A shows pollution, left to itself, slowly dispersing along the gradient to where the lower part of the water-table joins the river.

— To counterbalance this natural flow (Figure 5.18b) a pump can be installed at the level of the bottom of the contaminated zone; besides extracting contaminated water of the subsoil, pumping will also create a depression around the pump—the term 'cone of depression' is used—hindering the natural flow downstream; as a result, the pollution will be restricted to near its source and its dispersion outside that zone greatly reduced.

The characteristics of pumping and the cone of depression created can be estimated by modelling the system based on its hydrogeological and hydrodynamic data, type and behaviour of contaminants....

Treatment of the pumped water depends on the quantity of contaminant contained; in the best case a pump can be installed at sufficient depth so that the contaminated surficial part of the groundwater is not induced into a cone of depression which 'inhibits' the natural process of dispersion. Success of the operation rests on equilibrium between minimum pumping and least possible dispersion. Needless to say, the performance will be better with contaminating substances of low mobility.

The main advantage of this method lies in the relative ease of its design and implementation, provided the design is based on the most accurate

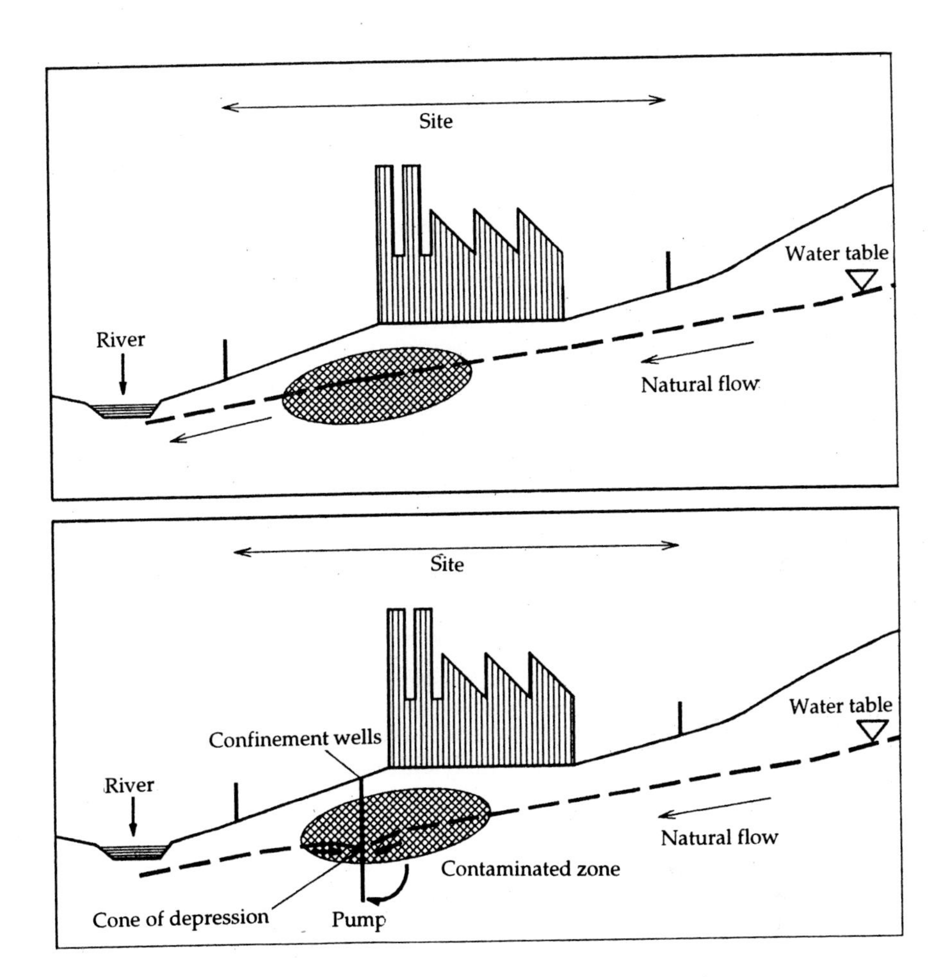

Fig. 5.18: Schematic sketch of a hydraulic trap assisting in the 'restriction' of a contamination under the site.

analysis possible of the characteristics of the aquifer. On the other hand, maintenance of equilibrium in the process over a long time, under conditions of eventual modifications of hydraulic conditions outside the trap, is difficult. Furthermore, since the contaminant is not destroyed, it may be obligatory to continue the operation until the entire stock of contaminants has been pumped out. Finally, it is evident that such a method requires a monitoring and follow-up of the changes in water quality through a network of piezometers.

5.4.3 Chemical Methods

This group of methods of treatment includes an ensemble of diverse techniques which can be classified by the type of reaction—this word used in a broad sense—that results.

First of all, one can distinguish the **methods depending on 'washing'** whereby the contaminating substances are mobilised or extracted and subsequently entrained in a transporting phase (most often liquid) to take them outside the medium they pollute. For example, washing with a detergent or acid solution.

This group also includes **more reactive processes** in which the pollutants are chemically transformed into other compounds, generally less complex and always non-contaminating, such as destruction of chlorocompounds by dehalogenation.

Finally, we shall also consider in this group electrochemical techniques in the broad sense, which involve electric energy to initiate the reaction, such as electrolysis.

Some methods are thus, strictly speaking, mobilisers of the contaminant entrained by them in another medium without destroying it (washing), very often by concentrating it in a phase where it will be treated at the appropriate time. On the other hand, some methods effectively destroy the pollutants during treatment (chemical reaction), transforming them into harmless substances.

In terms of application, on-site treatment of excavated earth or pumped water by chemical methods is preferable although sometimes this technology is adapted to in-situ application also. For electrochemical methods on the other hand, in-situ treatment is usually utilised.

We shall now discuss these methods in the following order:

- **methods of mobilisation and extraction,**
- **methods involving destruction by reaction,**
- **electrochemical methods.**

Mobilisation and extraction

Ordinarily similar to washing, simple or with detergent, mobilisation of a contaminant by a solution and its removal away from the contaminated site, offers several alternatives for almost all contaminants, depending on the type of mobilising agent used in the solution. The technique involving washing may be considered for either in-situ or on-site application.

In-situ application

The operation is simple: the soil is sprayed with a solution that infiltrates it and mobilises the contaminant. The solution containing the mobilised contaminant is pumped out through wells or through underground drains located on the downstream boundary of the concentrated contamination (Fig. 5.19). The recovered solution is purified and recycled for another spraying; the cycle may operate continuously for very long

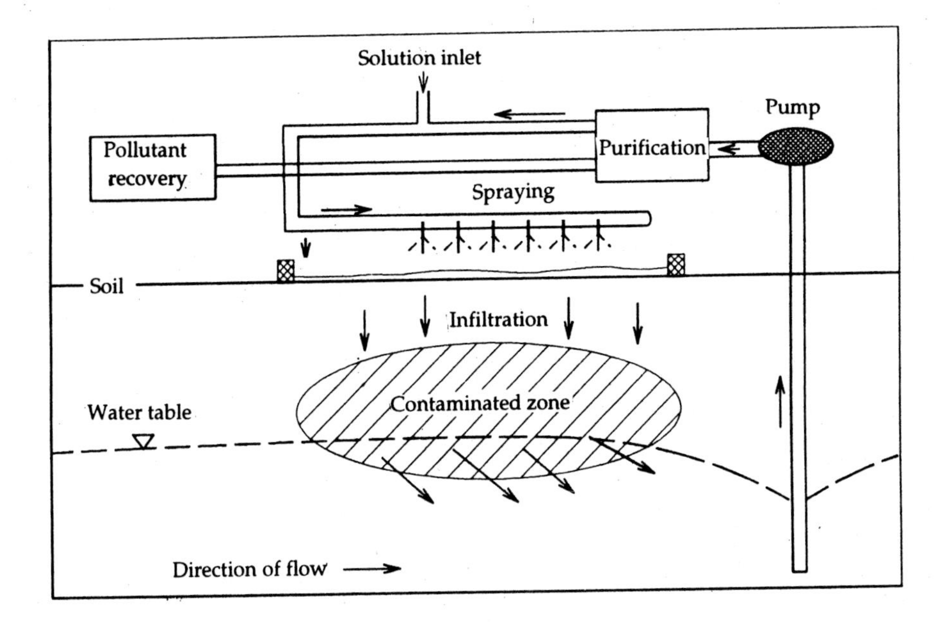

Fig. 5.19: Schematic sketch of in-situ washing.

periods (months or even years), with new doses of contaminated solutions brought in as the purified solution is pumped out. The main difficulty lies in close follow-up and managing to preclude migration of future contamination outside the recovery zone under the effect of penetration of the washing solution. Hence a thorough knowledge of the hydrogeological conditions and expertise in handling the hydraulic system are essential for success of the operation.

The advantages of the method and its application are: ease of implementation and low cost (provided the detergent is not costly) and its good performance in case of permeable soils. While the technique of washing performs well in sand or gravel, it becomes inoperative in finely granulated soils, such as silt, clay, clayey sand... because the medium rapidly gets choked with entrained particles under the effect of infiltration. Another disadvantage is that washing with reactive chemical solutions (acids) or toxic solutions (certain solvents) together with the effect of more or less pronounced choking (reduction in permeability) can result in marked irreversible alterations in the state of the soil after treatment. Finally, let us recall the risk of uncontrolled migration of the mobilised contaminants, laterally or towards the deeper layers of the aquifer and the consequent need for a recovery pump of adequate capacity coupled to a good network of control by piezometers.

On-site application of washing

The equipment used on site for washing contaminated soil or water greatly vary—each operator tends to develop and patent his own system, always based on the same principle of functioning and corresponding to the same contingencies of treatment.

In the case of a soil to be treated, the excavated material is first sieved to eliminate stones and coarse lumps, so that it can be introduced into the washing machine (a box with an agitator, rotating drum...). It is then mixed with a specified amount of the washing solution usually injected under pressure in the machine and agitated for some time to ensure a perfect soil-solution contact, resulting in mobilisation and washing of the pollutants (Fig. 5.20). At the outlet of the machine the cleaned earth is separated from the solution which carries, besides the contaminant, a large portion of the fine soil fraction. The mixture is separated by straining and/or centrifugation after the requisite period of contact and washing. (Usually a ratio of 10 parts, by volume of solution to 1 part earth is employed.) The fine fraction separated by physical procedures is dried and can be mixed with the coarser part of the soil to be relaid at site or properly stored. Treatment can be envisaged as several cycles of washing or a series of machines which treat the different fractions of soil separately with a solution appropriate for that fraction as explained below. After separation of the fines, the washing water is treated serially in steps designed according to the contaminants to be recovered (for example, oil-water separation, distillation, precipitation of metals in the form of salts...).

Groundwater pumped out is treated in a similar manner but in equipment adapted to liquids.

In many cases the contaminants are absorbed by the fine soil particles, clays and silts, or the organic matter. A first treatment thus consists of separating, by simply wet-straining, the medium and coarse fractions from the finer ones. The finer fractions carry the largest part of the contamination, sometimes in large concentrations, needing to be treated by more specific washing agents. It is thought that for a fraction larger than 2 mm a simple wash with water suffices to remediate the material by more than 95%. For medium fractions, some contaminants, not very soluble in water, require the addition of mobilising agents (such as surfactants for example), while for the finest fractions (< 0.2 mm) a specific treatment may have to be administered eventually by a procedure that differs from a simple chemical wash.

Regardless of whether the washing is done on site or in situ, mobilisation can be achieved in several ways—either by formation of an emulsion, dissolution in a solution or even by chemical transformation. Depending on the type of contaminant to be treated, as well as the technique to be applied, the solution used will differ. Several types of substances are used by

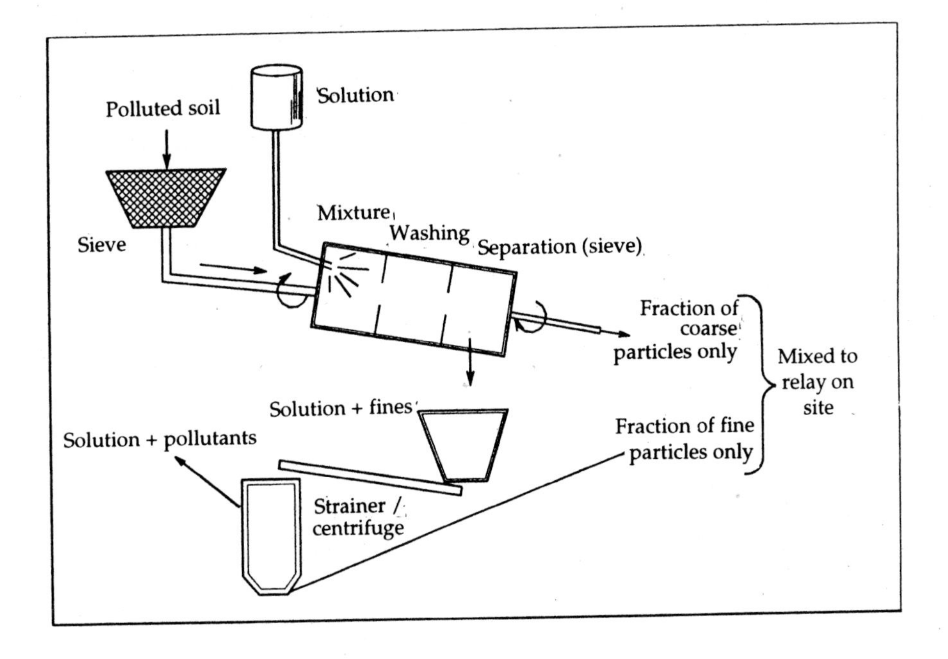

Fig. 5.20: Schematic sketch of on-site washing.

the operators, amongst which pure water, acid solutions, solvents (organic or inorganic) or surfactants may be specially mentioned.

Pure water, a less expensive solution posing no risk of degradation of the medium, can be used only for water-soluble contaminants. In many cases, however, water alone does not serve the purpose and additives are required. In some cases treatment by water can be improved by use of ultrasonic waves which act as catalysts and improve the washing of the material by oxidation of the contaminants (PCB, pesticides, PCP...) by the ozone generated.

Surfactants, whose primary function is to facilitate break-up of large organic molecules, are used for naturally hydrophobic substances. This applies in particular to groups of organic solvents containing halogens, aliphatic and aromatic hydrocarbons, either lighter ('LNAPL') or heavier ('DNAPL') than water. The technique of mobilisation by surfactants has been especially used for PAH, PCBs, semi-volatile chloro-solvents, asphalt and creosotes; it is suitable for organic contamination emanating from old coking units, wood-processing plants and even petroleum and petrochemical complexes. After mobilisation and removal of contaminants, the surfactant phase is 'cleansed': the volatile substances are removed by air stripping (see methods by evacuation) and the non-volatile extracts by solvent-extraction. The surfactant is subsequently concentrated by ultrafiltration in order to maximally reduce its dispersion in the aqueous phase. For in-situ treatments, anionic surfactants (for example, sodium dodecyl-sulphate) are preferred to cationic surfactants because they are not as readily adsorbed on the surface of soil particles. Generally speaking, they should, in this particular case, be non-toxic (or the least toxic!), biodegradable and, of course, less expensive. The last requirement should be fulfilled in the case of on-site treatments also. Table 5.6 lists the large families of surfactants—and their main characteristics—which are likely to be commonly used in remediation operations.

Table 5.6: Surfactants: families and characteristics (adapted from Anon., *The Hazardous Waste Consultant*, Sept/Oct, 1991)

Type	Surfactant	Use	Solubility	Reactivity
Anionic	Carboxyl acid salt, carboxyl	Good detergent	Water soluble	Tolerant to electrolytes
	Ester of sulphuric acid	Good wetting agent	Soluble in polar organic compounds	Sensitive to electrolytes
	Ester of phosphoric acid	Strong reducer of tensions		Resistant to biodegradation
	Sulphonate salt	Good emulsifier of oil in water		Resistant to acidic or alkaline hydrolysis
Cationic	Amino series	Emulsifying agent	Water solubility variable or low	Stable to acids
Non-ionic	Polyoxyethylic alkylphenol	Emulsifying agent	Water soluble	Good chemical stability
	Polyoxyethylic alcohol	Detergent	Non-soluble in water	Resistant to biodegradation
	Polyoxyethylic glycol	Wetting agent		Relatively non-toxic
Amphoteric	Sensitive to pH	Solubilising agent	Variable solubility depending on pH	Non-toxic
	Insensitive to pH	Wetting agent		Tolerant to electrolysis

Once the contaminant is mobilised, it is often useful to maintain it in the solution; this is the role of surfactants that prevent or restrict reattachment of the contaminant to the medium after its mobilisation.

Organic solvents can be used in certain special applications carried out in reactors on the site; it is, as a matter of fact, totally inadvisable to use this type of extraction for an in-situ application. Organic solvents for washing should be used only for specific applications, after being fully tested beforehand in the laboratory and adapted to the site conditions, case by case. The solvent is regenerated at the end of the extraction cycle by distillation and the contaminants concentrated and stored in containers. Major families of organic substances relevant to this extraction process are PCBs, PCPs, PAHs, pesticides, heavy hydrocarbons and asphalt. Thus, for example, such extraction may use pure propane, mixed in the reactor with contaminated materials (soils, wastes, water or mud) in order to extract contaminants of these groups from them. After separation of the propane-material phases, propane is vaporised and the contaminants concentrated in a residual liquid phase; the propane is subsequently recycled. To treat liquid effluents or contaminated water, propane is replaced with carbon dioxide. Developed on a commercial scale, the efficacy of this procedure is generally 90%. In a concrete case of treatment through extraction with propane carried out in Texas (Anon., 1991, op. cit.) in a unit producing creosote, efficacy of the order of 85% was obtained in the recovery of PCP and 95% for APH (see Table 5.7).

Table 5.7: Efficacy of extraction with propane of 'creosote' type substances (Anon., 1991, op. cit.)

Compound	Untreated soil	Treated soil	% Recovery
Anthracene	210/330	9	96/97
Benzopyrone	24/48	11/12	52/76
Benzofluoran	24/51	10/13	52/77
Fluoranthene	270/360	11	96/97
Naphthalene	69/140	1.5	98/99
Total PAH	2124/2879	110/123	96
Pentachlorophenol (concentrations in mg/kg)	210/380	52/58	85

Another application of extraction by solvents was accomplished by using amines dissolved in water to extract hydrocarbons and other organic compounds from soils or wastes. The procedure is carried out in cold conditions, given the excellent solubility of amines, but an atmosphere of nitrogen with complete absence of oxygen must be maintained (otherwise, there will be risk of explosion).

Supercritical extraction is a remediation procedure developed recently at the pilot scale (Jean, 1997). It is based on the very large solvent capability of a fluid in supercritical state, intermediate between the gaseous and liquid phases and characterised by a slightly increased viscosity and highly increased coefficient of diffusion compared to ordinary liquid solvents. The most commonly used* supercritical fluid is CO_2, mainly because of its non-

*Other supercritical fluids, such as H_2O, propane, butane etc. can also be used.

toxicity and low cost. To attain this state, the operation must be carried out under high pressure (higher than 200 bar) and low temperature (ten degrees or so) so that numerous organic molecules are solubilised. CO_2 is a non-polar solvent with which other co-solvents or complex organic solvents can be mixed to improve the efficiency of extraction, depending on the type of pollutant being treated. Although this procedure for the 'washing' of organic molecules is still at the pre-industrial stage, research is already underway for using it for mineral substances. Its cost is still too high to be really competitive.

More commonly, operations of extraction by acidic or basic solutions have been routinely successful and applied on site as well as in situ.

Extraction by acids is directed towards heavy metals, using primarily HCl, HNO_3 or H_2SO_4.

For bases, the extracting chemical used is essentially caustic soda ($NaOH$) and the contaminants to be removed may be cyanides, certain metals, as well as organic compounds such as amines, ethers or phenols.

A concrete example of in-situ extraction with acid for a Dutch site contaminated with cadmium (Cd) was published by Urlings (1990). On the site of an old factory producing photographic paper, about 30,000 m^3 of soil, representing a surface area of 6000 m^2, had been contaminated with Cd, the concentration varying from 5 to more than 20 mg/kg (with the background concentration less than 1 mg). The terrain was sufficiently permeable and characterised by a grain size ranging from 60 to 500 microns; it was impregnated with weak hydrochloric acid by surface spraying (Fig. 5.21). As the acid penetrated the soil, it desorbed the cadmium present and formed a solution. The acidic solution containing the metal was then recovered through a system of drains by pumping, treated on the surface to extract the cadmium and recycled. After one year of treatment, the concentration of cadmium in the soil returned to values less than 1 mg/kg. Treatment of the pumped effluent was accomplished by passing it over cationic resin.

Another application of in-situ extraction, carried out this time in an alkaline medium, may be illustrated by the treatment of deposits of earth or materials contaminated with cyanides. This application is significant because cyanides, noted for their ability to form stable complexes with metals, are widely used and have contaminated numerous industrial sites, metallurgical, mining and old gas factories. In the last case contamination arose as an undesirable by-product of gassification of coal.

Having spread in a natural medium, cyanides form very stable toxic complexes with iron (ferrocyanates) of bluish tinge (Prussian blue), which contaminate both surface and groundwaters by transformation to mobile ferricyanate. In oxidising and alkaline (pH > 8) conditions, the iron complexes are destabilised and the CN^- iron, which is soluble, can be mobilised and removed. Mixing cyanide-contaminated earth with a basic solution induces dissolution and washing of the contaminants from the medium. The recovered solution is then treated by acidification to precipitate the cyanide, concentrate it and subsequently alkalise it before reuse. The process can also be adapted for an in-situ treatment and the system was successfully used in Denmark (Pugholm, 1993) for treating, by spraying, heaps of waste containing sulphates and cyanides originating from an extractive industry. The 'leachates' were collected by a system of drains installed under and downstream of the heaps.

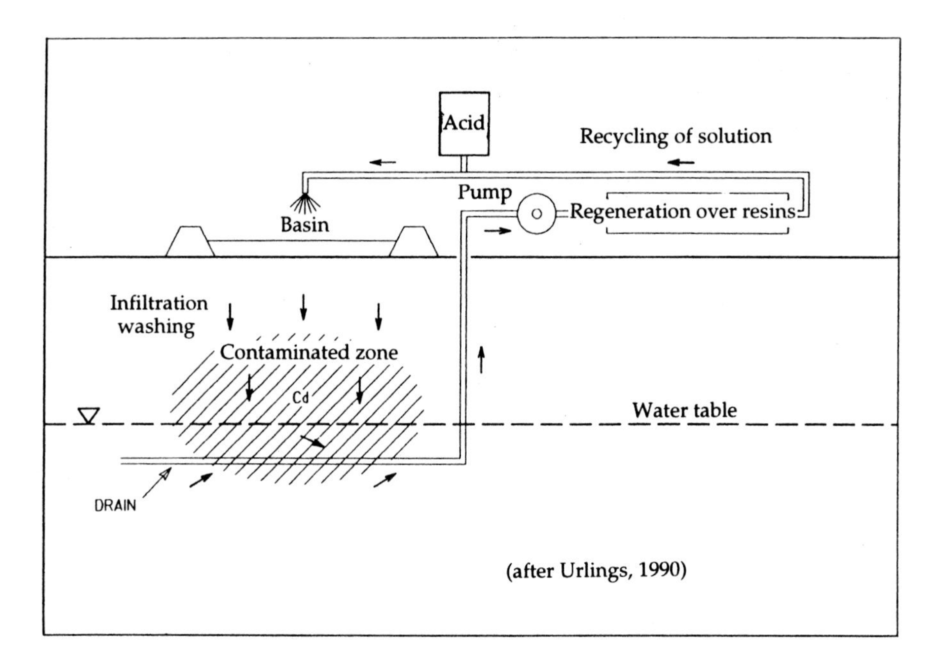

Fig. 5.21: Schematic sketch of in-situ mobilisation with hydrochloric acid and recovery of cadmium.

Chemical reactions

Another method of chemical treatment of soils or water consists of inducing a contaminant reaction through the use of certain reactants which generate compounds that are less toxic and eventually have a mobility different from that of the original contaminants, either higher so that products of reaction can be extracted, or lower so that a better immobilisation of the resultant compounds is obtained.

This process is generally limited to organic substances that are not very volatile and are difficult to biodegrade; it pertains thus primarily to chlorosolvents, PCBs, PAH and pesticides, but is also applicable to hydrocarbons and phenol compounds. In addition, some metals can be treated by this method, chromium for example.

This treatment is usually applied on site; in some cases in-situ applications have also been done.

Three types of major reactions are discernible:

— **oxidation,**

— **reduction,**

— **dechlorination.**

Each is briefly reviewed below. After a brief recapitulation of the theoretical basis, the mode of functioning and application in the domain of remediation, as well as the advantages and drawbacks are given and illustrative examples presented.

For a given organic compound, **oxidation** corresponds to a loss of electrons; the oxidation number of the oxidised product is higher than the initial contaminant while that of the added reactant decreases. This reaction corresponds to a natural phenomenon, commonly observed in soil chemistry. In the case of treatment of contamination, oxidation is induced by a powerful oxidising agent added to the material under treatment. In theory, although the potential oxidising agents are many, in practice only a few can be used because, on the one hand, some are very sensitive to pH conditions and, on the other, the form in which they are available is not necessarily conducive to large-scale treatment.

The two most commonly used oxidising agents are ozone and hydrogen peroxide.

Chlorine compounds such as ClO_2, ClO^- may also be mentioned in this context. The procedure is usually applied for treating cyanide pollutants and for the disinfection of water to make it fit for human consumption.

Ozone (O_3), an odourless gas characterised by a strong potential for oxidising, can be used as such for direct degradation of certain organic compounds, especially in the treatment of contaminated water or liquid effluents. It has the distinct property of non-storability as it quickly decomposes to oxygen (its half-life in water is only a few minutes); it thus must be produced on the site of utilisation, directly upstream of the treatment apparatus. Apart from its oxidising action, ozone can also be used to increase the concentration of oxygen in the medium and thereby facilitate biodegradation due to better bacterial growth (see Section on biotreatments).

A standard example of oxidation with ozone is treatment of water contaminated with hydrocarbons, wherein the specific action consists of diminishing the quantity of organic carbon dissolved in water through formation of CO_2.

Hydrogen peroxide (H_2O_2) is a more powerful oxidising agent than ozone: it can act at various levels. First, it can be used for direct degradation of complex organic compounds which are resistant to biodegradation (PCB, PAH, PCP...); next, it can modify the mobility of some metals; and finally, like ozone, it induces augmentation of the quantum of dissolved oxygen in the medium, thereby facilitating bacterial activity and biodegradation.

Its effect can be catalysed by the action of ultraviolet rays. Thus the process of oxidation of peroxide consists of adding H_2O_2 to the contaminated water and circulating it in a series of compartments under intense UV radiation. This will catalyse the reaction of oxidation of organic compounds present in water by activating the decomposition of the peroxide in radical OH, which reacts strongly with the organic molecules and thereby provokes their decomposition into harmless subproducts (H_2O, CO_2 and salts). The technique is mainly applied to effluents contaminated by PCBs, chloro-solvents or hydrocarbons.

Generally speaking, treatment by oxidation by adding ozone or peroxide seems to be very effective for degradation of appropriate contaminating substances with the additional benefit of augmenting the total concentration of dissolved oxygen in the medium and, consequently, the biodegradative action of bacterial flora.

However, the process is not specific, which means that, depending on the composition of the soil or other compounds present in water, part of the added oxidising agent may be 'wasted' in oxidation reactions other than the one sought for degradation of a particular contaminant. For example, in the presence of Fe^{2+} (not well-aerated medium, rich in organic matter), the oxidising agent will preferably transform ferrous iron (Fe^{2+}) to ferric iron (Fe^{3+}).

Finally, another disadvantage of oxidation (especially by ozone) concerns the case when degradation of pesticides is desired. In fact, in the case of some pesticides, oxidation transforms the original product to a similar one, also toxic and additionally much more stable in terms of time; such is the case of transformation of aldrine to dieldrine, which degrades much more slowly.

By definition **reduction** corresponds to the reaction inverse of oxidation, i.e., the state of oxidation of a substance has to be reduced (or its number of electrons augmented) by adding a reducing agent or donor of electrons. It also includes a natural phenomenon, well known in physical chemistry of soils, namely, that the addition of a reducing agent merely accelerates the process. Some compounds are more prone to reduction than others, inasmuch as they are effective acceptors of electrons. This type of reduction relates in particular to some organic contaminants (aromatics, pesticides...) and some metals (chromium and selenium).

To degrade organic compounds by reduction, the main agents used are metals, added in pulverised form, which act as catalysts. The one most often employed is undoubtedly iron, given its easy availability and low cost.

Reduction of an organic compound implies several types of reactions, such as dehalogenation or saturation of aromatic molecular cycles, and in each case aims at transformation of a toxic substance into a harmless one. Thus one may consider the example of transformation of DDT into DDA, according to the equation:

$$\mathrm{Fe + H_2O + Radical\text{–}Cl \Leftrightarrow Fe^{2+} + OH^- + Cl^- + Radical\text{–}H},$$

or even the transformation of chlorobenzene to cyclohexanol, according to the equation:

$$\mathrm{Fe + 2H_2O + R\text{–}CH{=}CH\text{–}R \Leftrightarrow Fe^{2+} + 2OH^- + R\text{–}CH_2\text{–}CH_2\text{–}R}.$$

This type of reaction normally requires quite strict pH conditions (pH of 6 to 8), which must be maintained throughout the treatment.

For metals, an important application of degradation by reduction pertains to changing the valency state of chromium. The two stable states of oxidation of chromium naturally present in the medium are Cr^{3+} and Cr^{6+} (Richard and Bourg, 1991). The first, present in mineral phases of oxides-hydroxides, is not very mobile and is non-toxic. It can be mobilised only in conditions of acidity and oxidation. The second, contrarily, is mobile and very toxic and is present in the form of a chromate ion in natural water.

Used in the steel industry and for surface treatment as well as in tanneries and the chemical industry, chromium can be a cumbersome contaminant of the natural environment. In its toxic form, chromium becomes rapidly dangerous for the ecology and human health; it is recommended that its concentration in natural water should never exceed 50 microgrammes per litre.

When a pollution by Cr^{6+} reaches the water table, its 'in-situ' immobilisation in the non-toxic form Cr^{3+} represents a useful alternative to remediation (provided the water table is at small depth, a few metres below the surface). The operation consists of creating directly downstream of the sphere of dispersion of the pollutant such conditions of the medium as would provoke reduction of Cr^{6+} to Cr^{3+}, thereby immobilising the chromium in its inert form.

This type of 'chemical barrier' was tested at the pilot scale in the USA where a trench dug downstream of the flow of groundwater was filled with iron filings floating in an acidic solution (Fig. 5.22). When the contaminated groundwater crossed the trench a reduction reaction was triggered, transforming Cr^{6+} into Cr^{3+} by oxidation of ferrous iron, in accordance with the following equations:

$Fe + 2H^+$ (acidic) $\rightarrow Fe^{2+} + H_2$ gas (which escapes)

$Cr^{6+} + 3Fe^{2+} \rightarrow Cr^{3+} + 3Fe^{3+}$.

The chromium was thus literally stopped at the level of the trench in a harmless form and the remediated groundwater continued on its downstream path. During treatment the 'chemical barrier' must be constantly monitored to ensure maintenance of sufficiently acidic conditions, as these facilitate reduction of iron to ferrous iron.

The main advantages and drawbacks of treatment by reaction of reduction are identical to those of oxidation. Here, too, the conversion of compounds susceptible to quick reduction is high but again lacks specificity of reaction, i.e., other compounds naturally present in the medium will be

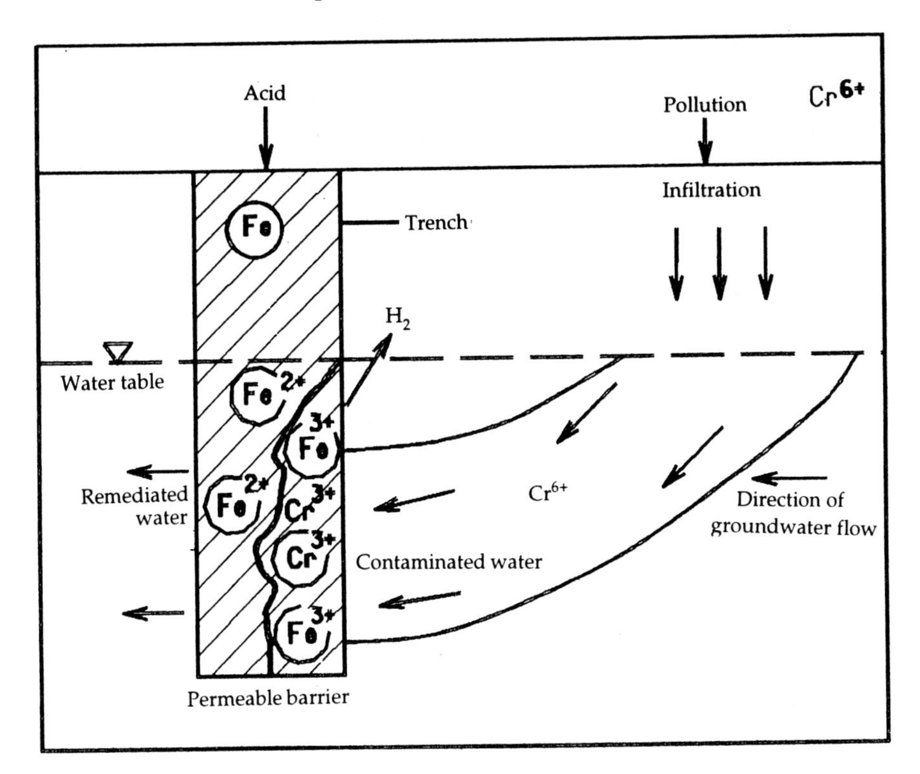

Fig. 5.22: Schematic sketch of chemical barrier for chromium, with reduction of Cr^{6+} to Cr^{3+}.

affected by the reduction process along with the contaminant to be degraded, implying some 'wastage' of the reducing agent.

Dechlorination is a treatment wherein the objective is to transform higher chlorine-content molecules to lower chlorine-content molecules. In principle, the higher the number of chlorine atoms in a molecule, the more toxic the compound. The purpose of dechlorination is to obtain molecules that are less toxic and usually more soluble in water. The technique of dechlorination was specially designed for PCBs, PCPs and chloro-solvents. Several procedures are being marketed today, some operative at the field-scale, others still at the pilot stage.

The general principle consists of replacing, entirely or partially, the Cl^- ions in the organic compounds by OH^- radicals. Two different chemical reactions take place in two distinct variations of the procedure. In the first, the contaminated material is mixed with sodium bicarbonate, often in the presence of an organic catalyst (BCD or 'base catalysed decomposition' procedure); the chlorine compounds are decomposed, (partially) volatilised and recovered. In the second, the dehalogenation agent is an alkaline glycolate polyethylene (mostly of potassium) and the Cl^- radicals are replaced by glycol groups and the toxicity of the original molecule is thereby significantly reduced.

The device used for implementation of the procedure is the standard one used in on-site chemical treatments (Fig. 5.23): the excavated soil is crushed and sieved before being introduced in the reactor in which the reactants have been poured earlier. The mixture is brought to the temperature of reaction and the liberated gases recovered and directed towards the unit specified for gas treatment. At the end of the process the remediated soil is taken out and stored.

When applied to the treatment of PCBs, this procedure attains an efficacy of more than 90% in degradation; similarly, in the case of wood-seasoning installations, the results obtained have been 95% destruction for PCPs but only 60% for PAH. The procedure is currently being used at the operational scale, particularly in the USA.

Electrochemical techniques

Used for the last fifty years or so for cleaning soils or assisting in the recovery of petrol, electrophoresis is based on the application of a continuous current in a terrain, inducing migration of positive ions towards the cathode through the aqueous solution in the soil.

In the last few years the technology has been developed and applied for remediation of polluted soils and is known as 'electrorehabilitation' of soils.

Figure 5.24 presents a schematic sketch of its theoretical functioning: two electrodes are installed in the soil to be treated, by means of which a very strong current is circulated. The potential difference induces in the

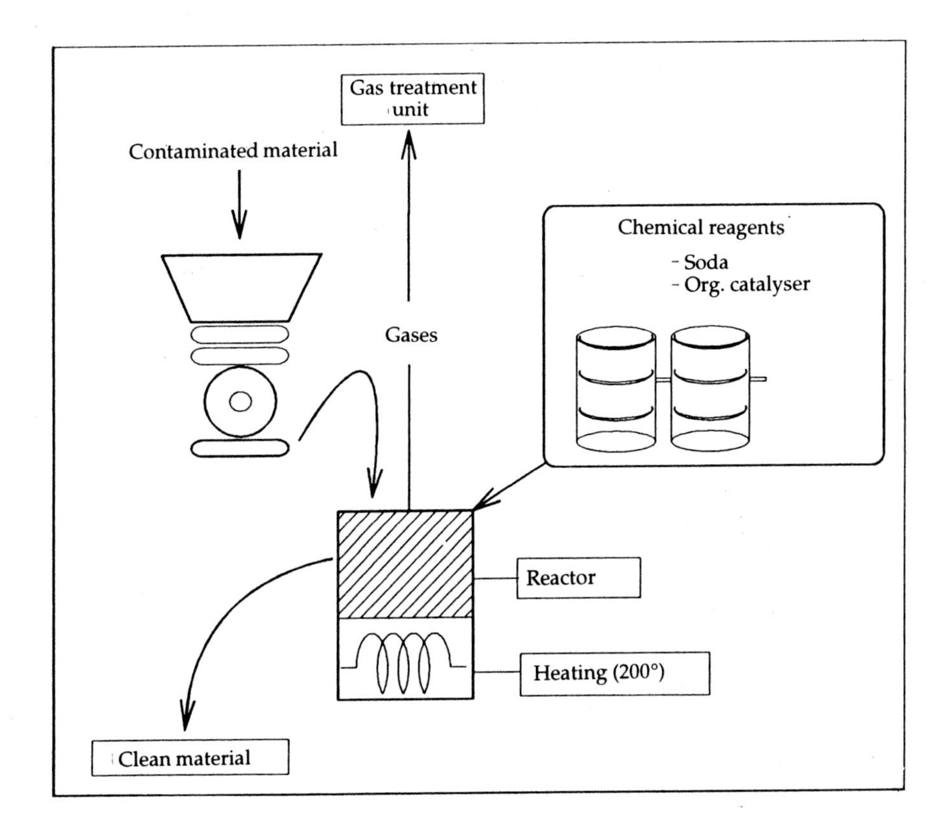

Fig. 5.23: Schematic sketch of the device used for dechlorination with caustic soda (also called the 'BCD' system).

medium, ionisation of part of the compounds present and their migration towards the electrodes via the soil-water. By definition, the positive metallic cations concentrate at the cathode (Cu^{2+}, Zn^{2+}, Cd^{2+}, Pb^{2+} ...) and the anions (SO_4^{2-}, Cl^-, $CN^=$, OH^-...) at the anode. While crossing the (semi-permeable) membranes of the electrodes, the ions are extracted from the soil and transported by the intermediary of a solution towards a purification apparatus in which the compounds recovered are extracted and concentrated, after which the transporting solution is recycled. The solution is made up of water and chemical additives, such as HCl (for the cathode), NaOH (for the anode) or of complex agents. The physicochemical characteristics of the medium (pH, EC, concentration of certain ions, carbonates...) are regulated by the intermediary of the solution circulating in the electrodes, to prevent precipitation of salts at the electrodes and, consequently, clogging of the system; the term electromigration is also used.

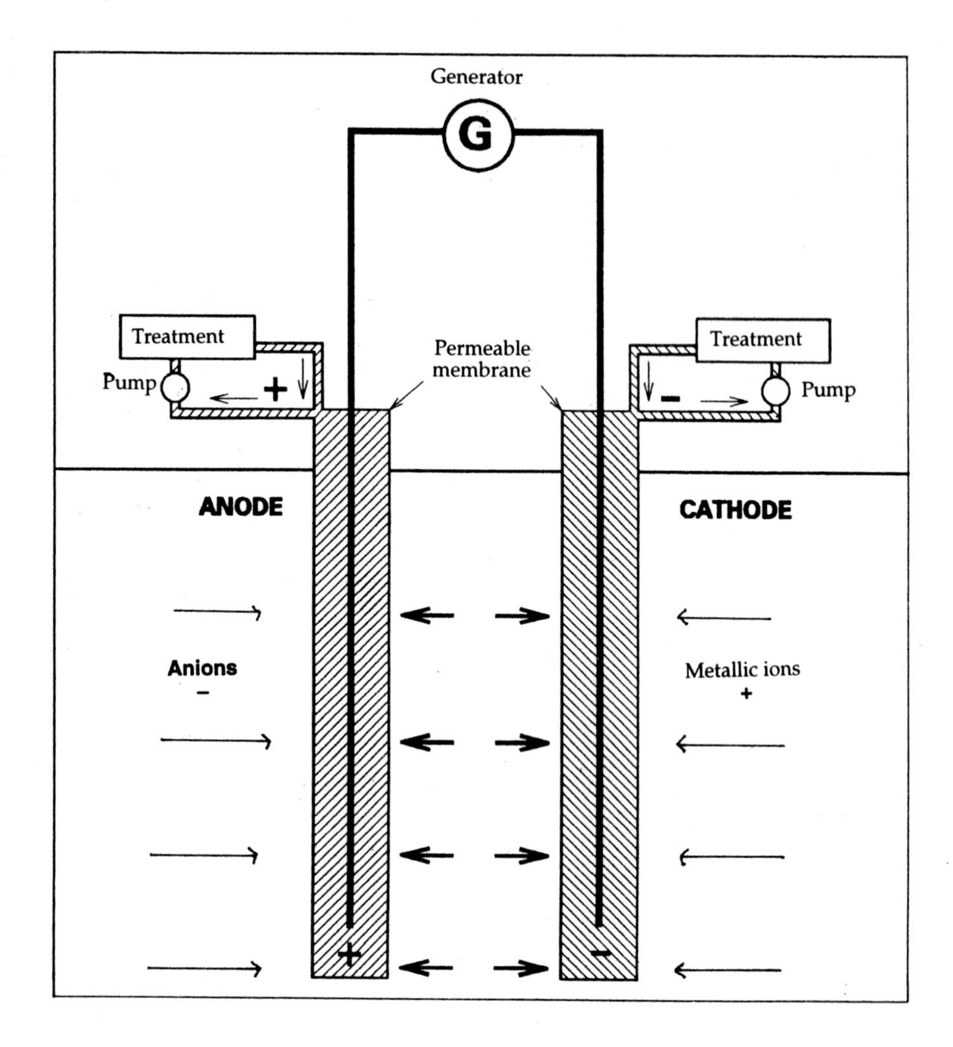

Fig. 5.24: Schematic sketch of an electrorehabilitation system.

The procedure handles all ionisable compounds, in particular nitrates, phosphates, cyanides, polar organic compounds and especially heavy metals, for which to date it has been employed more frequently. By this method the soil and groundwater can be treated simultaneously; it can only be used in media saturated with water. It is particularly efficient in porous but not very permeable media, i.e., media in which the grain size of the substratum varies from fine sand to clay. This characteristic constitutes an important advantage over the standard techniques (pumping, venting, biodegradation, dissolution in water...), whose implementation is often difficult when the permeability of the medium is low.

The technique can be applied in situ and also on excavated earth gathered in heaps and sufficiently humidified; whatever the case, the electrodes, regularly spaced in tandem, are driven in up to a certain depth, depending on the section of soil to be remediated and the thickness of the heaps in which the earth is stacked. Cathodes and anodes follow each other in parallel lines and the distance between them is calculated in terms of the physicochemical parameters of the terrain.

The method can also be applied in isolated electrical containers, with fixed electrodes, in which the material to be remediated is treated. Design and dimensioning of the device can be modelled in terms of the characteristics of the terrain, hydrological (velocity of underground water, porosity, permeability...) and electrochemical (electrochemical mobility, base-exchange-capacity...) as well as in terms of the technical data of the electrical system to be installed (radius of electrodes, configuration, voltage and current applied...).

The efficiency of the system is high depending on the electrical energy consumed and the time for which the operation lasts. Of course, the type of contaminant and the form in which it is found are determinative for the result obtained; the pH of the soil and its ion-exchange-capacity (Lageman, 1993) are likewise important. Recent experiments showed that the efficacy can be significantly augmented if the soil is acidified beforehand, or mixed with biodegradable complexing agents. This author cites an efficiency of 30 to 90% for such metals as Zn, Pb, Cs, As, or Cu, corresponding to a treatment of 1.5 months to 2 years' duration with initial concentrations of several thousand mg/kg in the soil. Consumption of electrical energy was high, of the order of 40 to 400 kWh/ton, and the volume treated varied from several tens to more than 7000 m^3.

Finally, another useful application of electrorehabilitation consists of establishing an electric barrier at the level of the water table by a series of electrode-pairs (cathode-anode) in tandem, driven in the soil to the level of the phreatic surface, perpendicular to the axis of water flow and thus the flux of contamination. The current directed into the soil creates an electric field around the electrodes, preventing downstream migration of the contamination dissolved by the groundwater flow. Further, at the present experimental stage, this technology combines the effects of chemical confinement and recovery of contaminants by electrical means.

In a lighter vein, mention may be made of the 'acoustic treatment' which is still at the experimental stage and is not, in the usual sense, a procedure by itself but is used in complement with the method of 'flushing' (in-situ washing). It is based on the fact that the vibrations created by acoustic waves can facilitate the mobilisation of non-volatile organic pollutants. As a matter of fact, during the pollution of soil by hydrophobic compounds, such as residual mineral oil, a part of the pollution is trapped in the pores of the soil in the form of pure substance and droplets of the pollutant are retained in the saturated zone by capillary forces. To compensate these forces and facilitate the mobilisation of the substances an additional force is required. Sound waves are introduced into the soil matrix, generating vibrations in the medium. At the frequency of resonance, the drops of oil subjected to these acoustic waves break up into smaller droplets which can be more easily eliminated by the technique of 'soil flushing'. Tests on columns of polluted sand confirmed the high efficiency of washing accompanied by subjecting the column to a flux of acoustic waves.

5.4.4 Thermal Methods

Commonly used today, thermal methods represent a major option amongst the plans of remediation. Thus up to 1992 incineration was the method selected for treating contaminated materials on nearly one-third of the sites remediated in the USA (EPA, 1993b).

Thermal technologies are based on the same principle, namely, heating the contaminated material to extract the pollutants and, in most cases, concomitantly destroying them. But though the principle appears simple, these techniques require a high degree of technical competence, whose acquisition and implementation remains complex.

These methods can be applied on site or away from the site, which is more often the case. In-situ methods are presently still at the experimental stage.

We shall describe below, in this order:

— **techniques of incineration,**

— **thermal desorption,**

— **pyrolysis**

— **vitrification** (in terms of results, this is similar to the methods of stabilisation described earlier in the section on techniques of 'trapping').

Incineration

Incineration is a very old technique skillfully used by man down through the ages to get rid of wastes. During the last fifty years or so it has undergone considerable technical development, enabling disposal not only of all types of wastes and families of contaminating substances, but also more recently treatment of soils and other contaminated solid products of the natural milieu.

Incineration uses high temperatures to destroy contaminating substances, which are converted into carbonic gas and steam, leaving behind various other products of combustion. This method is very effective for a very wide range of substances derived from organic compounds; it cannot, however, destroy metals, which reappear either in the gaseous emissions when volatisable, or in the solid residues of combustion (ashes).

Transforming organic compounds into harmless simple molecules, incineration may be considered one of the exclusive methods of remediation in which the contaminating substances are actually destroyed, unlike other numerous technologies which consist of operations whereby the pollutant is transferred to another place or is confined. So, it may be preferable to other non-destructive methods.

Technically, several systems exist, such as devices with a fluidised or circulating bed, infrared technique or a rotary kiln; the last is definitely the most common.

A simplified schematic sketch of incineration by a rotary kiln is depicted in Figure 5.25.

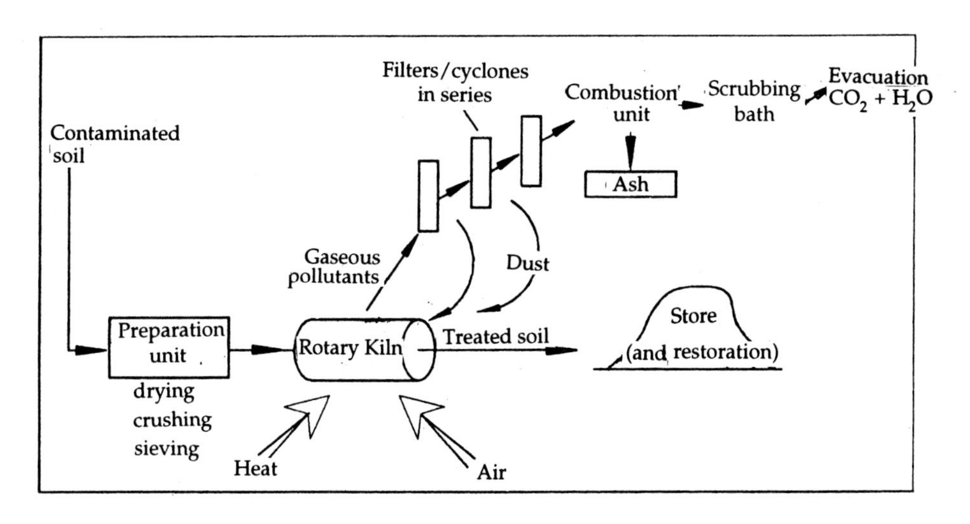

Fig. 5.25: Simplified schematic sketch of incineration in a rotary kiln.

The excavated soil is first dried, then crushed and sieved. Generally the coarsest elements, larger than a few mm, are not incinerated but sent to a refuse dump. If there is risk of evaporation of volatile contaminants, precautions have to be taken to recover the emanations, for example by operating a completely closed chamber in which a suction system collects all the gases which are then treated separately.

Incineration is generally carried out in two steps: volatilisation and destruction. Volatilisation is accomplished in a rotary kiln at a temperature of about 400°C and destruction in a combustion chamber at more than 1000°C.

After preparation, the material is placed in the rotary kiln, heated and stirred, and the volatilised gases carried away by an airflow. At this temperature all the pollutants present are volatilised and only the solid material left. The soil freed from the contaminating substances is removed from the kiln and can be restored after cooling. The gaseous flow, loaded with volatilised products, is directed into a combustion chamber of very high temperature (between 900 and 1300°C) in which all the organic compounds are destroyed, the molecules decomposing into CO_2 and H_2O.

Between the rotary kiln and the combustion chamber, the gaseous flow is passed through a series of cyclones and filters to retain the fine solid matter (dust) which had been entrained during the kiln treatment; subsequently, it is redirected towards the kiln or treated soil store.

Any chlorine or sulphur liberated from certain organic compounds (chloro-solvents, hydrocarbons...) is extracted from the gaseous flow by scrubbing it in an alkaline solution before discharge into the atmosphere.

Treatment of soils contaminated by PCBs requires special precautions to preclude formation of furans and dioxines; for nitrogen compounds also the technical characteristics of the procedure have to be adapted to prevent formation of N_2O.

When the soil to be treated contains heavy metals, the system of incineration has to be specifically adapted. Volatisable elements such as Zn, Cd or Pb, for examples, get entrained with the gases, are recovered, oxidised and concentrated separately, then sold to refineries. Non-volatile metals, such as Al or Fe, remain in the treated soil. Depending on the quantum at the exit and the norms to be followed, the soil may be stored, dispatched to a refuse dump or be subjected to another specific treatment.

A device for continuous control of the gaseous flow aids better management of the procedure and allows only non-toxic products (mainly CO_2 and steam) to escape into the atmosphere. Similarly, the composition of the solid products, soil, residues and ash, is analysed and checked at different points of the circuit. The efficiency obtained corresponds, in the best case, to 99.99% of extraction and destruction of all organic compounds initially present. The efficiency for removal of chlorine and sulphur from the gaseous flow at the exit attains similar levels.

Thermal desorption

For both volatile and non-volatile compounds, thermal desorption is a useful option and is easier to implement than incineration. The process comprises two steps: a unit of vaporisation of pollutants, of the type as used in incineration, and another unit of treatment of extracted gases.

The process is usually carried out in a drier—rotating or stationary; the dirty soil is heated and eventually removed, and the volatilised compounds are carried by a gaseous flow (air or nitrogen) towards the second unit. The working temperature is of the order of 250 to 450°C. The time of residence in the drier is adjusted according to the quantity of volatile contaminants contained and may vary from a few tens of minutes to several hours.

The gaseous flow carries the volatilised compounds towards the gas treatment unit, comprising a (or several) condenser(s). The gaseous flow first passes through water where it is partially cooled and solid particles are separated, which would otherwise remain entrained in it. In the condenser, the organic compounds are concentrated in liquid phase by cooling and separated from the gaseous phase: about 90% contaminants are arrested at this stage; the remainder is subsequently absorbed over activated carbon and the carrier gas then recycled and redirected to the drier. Solid particles are separated from the cooling water and returned to the drier while the water is recycled.

Results have shown nearly 95% efficiency of extraction of contaminants for initial concentration in the soil under treatment not exceeding 10%. Besides chloro-solvents and aromatic compounds, the technique can be used for pollution by PCBs also. The level of concentration obtained after treatment is in almost all cases less than that prescribed by norms in force or targets fixed. Thus, for example, for a specific case of treatment of polycyclic aromatic hydrocarbons, the initial concentration of the order of 500 to 2000 μg/kg fell to less than 400 μg/kg after treatment.

Pyrolysis

Pyrolysis consists of heating the polluted material in the absence of oxygens to a temperature of a few hundred degrees. In the procedure, there is no contact between the heating gas and the contaminated matrix. The pyrolysis causes 'cracking' of organic molecules into simple compounds, e.g. methane. The polluting material is thus transformed to gas leaving a solid residue containing ash and carbon ('coke'). At the final stage, the gaseous fraction to be treated is limited to organic constituents that can be recovered as a by-product or be adsorbed and a small fraction of inert gases maintained in circulation in the kiln. The temperature (450–600°C) used in this procedure is much lower than that used in incineration (900–1200°C). Due to the absence of oxidation and the relatively low temperature, this procedure is well suited for soils rich in organic matter and wastes. A constraining factor is the relative humidity which must be kept as low as possible. At present this technique is not being used in France.

Vitrification

This technical procedure consists of transforming by raising the temperature, the contaminated soil locally to an inert molten mass. It is thus an in-situ thermal method, resembling the techniques of in-situ heating and stabilisation (see the preceding group of techniques).

Fusion of soil is accomplished by driving electrodes in the soil to the depth necessary for treatment, and inducing through an electric current an intense rise in temperature, of the order of 2000°C (Fig. 5.26). A layer of sheet graphite and chipped glass is deposited on the surface between the electrodes (sketch 1 of the Figure) to activate the reaction in the soil which is a priori a bad conductor (especially when it is not very humid). At this temperature, the matrix of the soil is brought to fusion and is transformed by pyrolysis to silicated glass, in which all the compounds present are melted or vaporised.

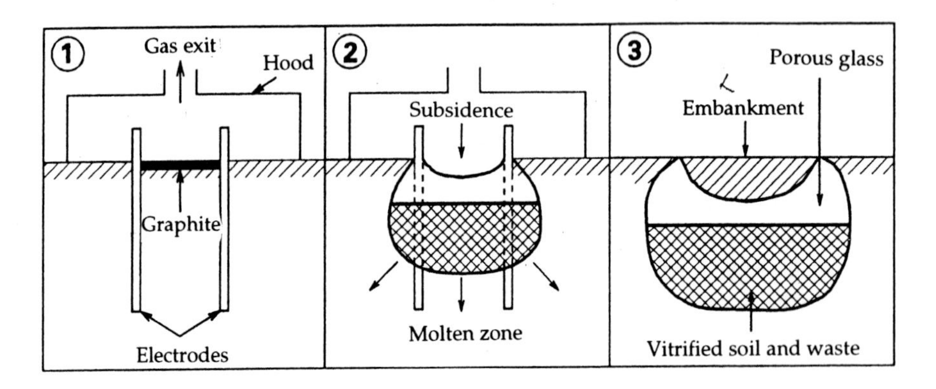

Fig. 5.26: Schematic sketches of in-situ vitrification (extract from Groo, 1987).

To facilitate collection of gases escaping during the reaction, a hermetic hood and a suction device are placed above the zone being vitrified (sketch 2); these gases are treated separately according to a specific plan. At the end of the reaction, and after cooling, the material is in stable form, chemically inert, without leachable elements, very similar in stability to granite for example (sketch 3). The organic compounds in a contaminated soil are thus volatilised while heavy metals get incorporated in the glass.

This technique, theoretically applicable to every type of contamination, is still at the laboratory stage (a few kilos of earth) or pilot studies of small zones (several to a few hundred tons) and not yet at the scale of an entire site. To the best of our knowledge, this method has yet to be tested in France.

Although this method has all the advantages of the in-situ methods (notably not involving excavation) of rendering the soil completely inert and very stable over time, it prevents, for the same reason, reappearance of a vegetal carpet on a material which is impermeable and glassified. Furthermore, it requires very large energy consumption (and hence is expensive—see later in the book); finally, the reaction at high temperature can provoke migration of the pollutants towards the exterior of the zone, thus extending the zone of dispersion.

5.4.5 Biological Methods

Principle

Biodegradation is a natural phenomenon. It is the result of degradation of organic molecules or minerals (nitrates) by micro-organisms (bacteria, fungi...). Their growth takes place through the intermediary reactions of oxidation of organics which provide the source of energy for the process. This process is also called 'growth of biomass'. This reaction also involves other elements, nitrogen and phosphorus which participate in the protein and DNA synthesis, and sometimes, also as oxidising agents. Nitrogen and phosphorus are called 'nutrients'. In conditions where air is available—the term aerobic is used—the role of oxidising agent is played by atmospheric oxygen; in conditions of absence of air or oxygen—anaeorobic—this role is played by nitrates, sulphates, or even oxygen-containing organics. The reaction is of the 'redox' type, which means that the carbon atoms lose electrons for the benefit of the oxidising agent, which acts as the 'acceptor of electrons'. The process develops according to a chain reaction wherein the carbon compounds are transformed, by successive break-up, into less and less complex molecules until simple products, generally CO_2 and H_2O, are obtained (Fig. 5.27). The products of transformation are termed metabolites.

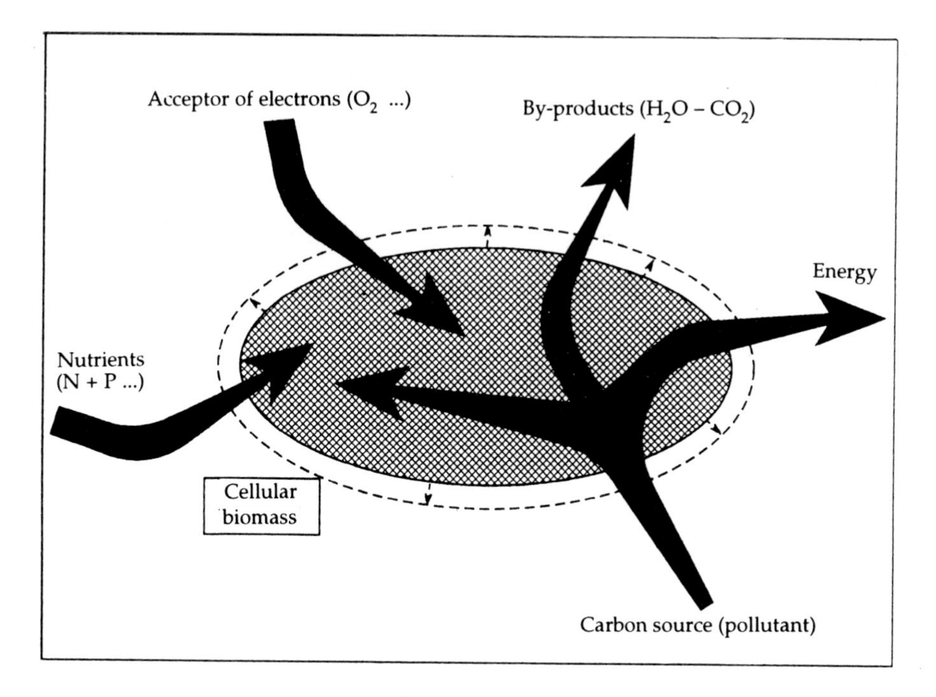

Fig. 5.27: Schematic representation of biosynthesis.

Remediation by biological means consists of stimulating this natural phenomenon to augment its efficacy, in order to destroy in minimal time the organic contaminant consumed as the source of carbon. Although the contaminant provides the necessary carbon, the quantities of nitrogen and phosphorus present in the medium under treatment may be too small. Balanced proportions of phosphorus/nitrogen, in relation to the quantity of carbon present, is necessary for success of the treatment. An adequate quantity of the oxidising agent has likewise to be introduced in the medium to be remediated.

Bioremediation is a method of actual remediation and not a simple transfer of the contaminant from one medium to another (as happens in the numerous methods described earlier), because the molecules are decomposed—and thus destroyed—in the course of the process. This technology is applicable to most organic compounds and today there are several micro-organism strains available capable of 'attacking' almost all organic and some mineral molecules.

Compared to other remediation techniques, bioremediation is now assuming greater importance; thus, in the USA its technological development is maximum and it is the second most used (after venting techniques) alternative to 'standard' solutions of confinement and incineration, for treating volatile and semi-volatile organic substances. The two graphs in Figure 5.28 show for the period June 1992 to April 1993, the distribution of sites treated in the USA by biodegradation according to type of contaminant and type of medium (EPA, 1993c). It can be seen from the second graph in the Figure that soil and groundwater predominate, in that order, and that in-situ operations prevail when the hydraulic conditions permit. The contaminants that predominate in the first graph are hydrocarbons constituting petroleum and products of treatment of wood, such as creosotes (group of phenols) or PAHs; PCBs and explosives are grouped in the category 'others'.

Groups of contaminating compounds which constitute 'good targets' for biodegradation, comprise;

— hydrocarbons constituting petroleum, such as gas oils, fuels, petrol, kerosene, mineral oils ... including the benzene group (benzene, toluene, xylenes...).

— wastes of production and processing of petroleum, mud, sludges and oil residues;

— organic products and residues of basic chemical industry, for example, alcohols, acetone, phenols, aldehydes and other solvents;

— halocarbon compounds, including aliphatic solvents (trichloroethylene, chloroform...) and aromatic solvents (chlorobenzenes) as well as PCBs;

— more complex compounds of polycyclic aromatic type (PAH), pesticides...;

— nitrates and sulphates.

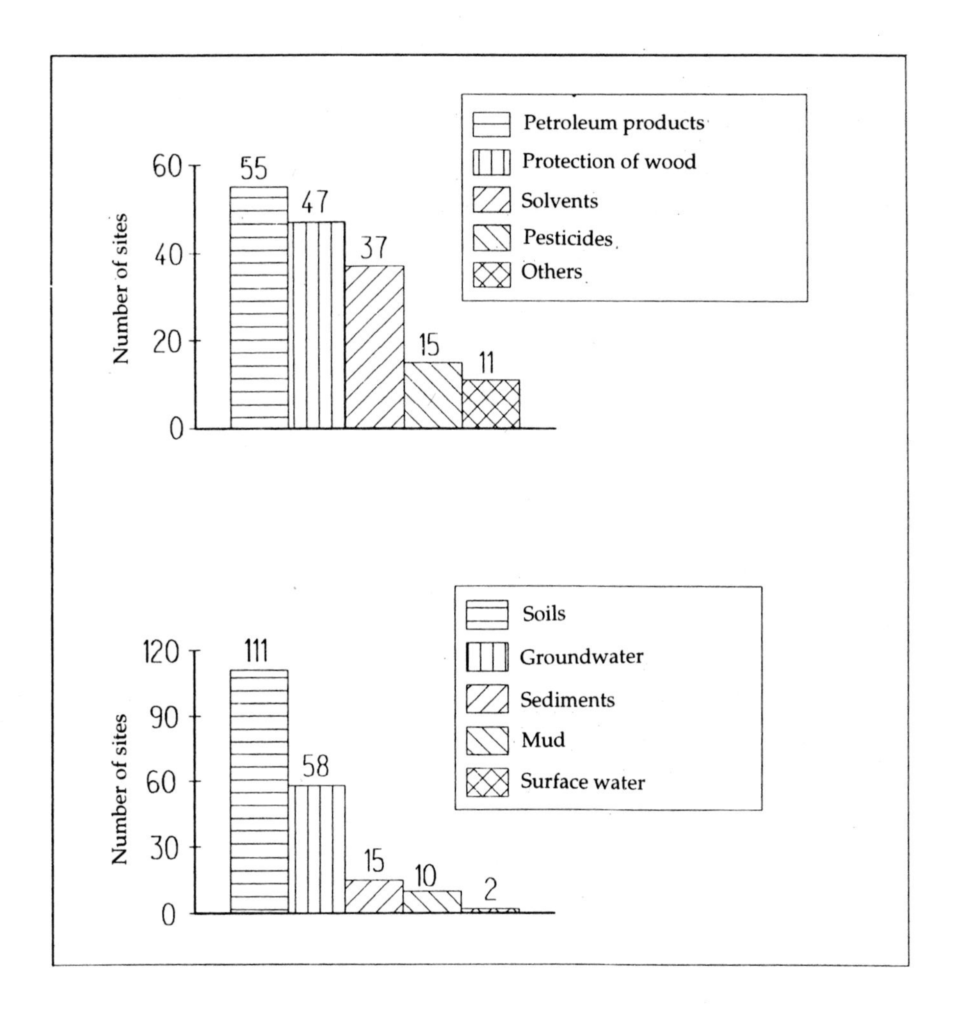

Fig. 5.28: Distribution of sites treated by biodegradation (159 sites, multiple choices possible; source; EPA, 1993c).

Application of biodegradation to inorganic compounds remains very limited and at the experimental stage, with nitrates an exception. As for heavy metals, biological means are commonly employed for fixing them in the given medium and thus preventing their migration or, contrarily, washing them out of the medium contaminated by them. This is described by the term bioleaching or biofixation and not biodegradation.

Whether for organic contaminants or some others, implementation of remediation by biological means must take into consideration two major requirements:

— materials added for treatment must be in contact with the contaminated medium insofar as possible;

— growth of biomass and transformation of contaminating compounds must be controlled and managed in a manner that takes into account the removal of by-products produced.

When these two requirements are met, remediation by biological means can be applied in two ways: either directly in situ by introducing into the subsoil the additives necessary for the growth of the biomass, or by on-site treatment of the water pumped or the soil excavated in specially equipped units. Over the last ten years or so, numerous methods have been developed for on-site or in-situ applications with specific adaptation to the type of contaminant and/or the medium under treatment.

We shall describe some methods, including those most commonly used and especially:

— ***bioreactors***

— ***biodegradation in pile, including composting, 'landfarming', 'biopiles' or 'biohills'***

— ***in-situ application by injection***

— ***'bioventing' and 'biosparging'***

— ***biobarriers and biological screens***

— ***phytoremediation***

— ***technique of 'humid zones'.***

Technological feasibility

The effectiveness of any biodegradation technique employed for decontaminating a particular site depends on numerous factors, which must be taken into account during design of the operation. In fact, there is no standard procedure which can be universally applied and whose success is guaranteed. Before proposing employment of bioremediation, a number of precautions must be exercised, lest there be risk of the method becoming totally inoperative; certain tests must be conducted and certain properties of the medium under treatment or the contaminant to be degraded measured.

• The first factor to be considered is the **biodegradability of the contaminant**, a concept difficult to define, sometimes not very precise, which depends not only on the substance to be degraded per se, but also on the other chemicals present and thermal conditions. Table 5.8 lists some compounds and their degradability 'status'; it can be seen that some may be considered biodegradable by one author and resistant to biological treatment by another.

Considered in isolation, a very large number of organic molecules are now considered biodegradable; on the other hand, when one is dealing with a mixture of products, phenomena of inhibition develop, halting the process of biodegradation, sometimes accompanied by poisoning of the

Table 5.8: Biodegradable nature of some organic compounds (extract from Lagny, 1991)

Compound	Biodegradability
Acetone	Degradable
Benzene	Degradable
Toluene	Degradable
Gas oil	Degradable
Phenol	Degradable
Trichloroethylene	Resistant
Chloroform	Resistant, degradable
Pentachlorophenol	Resistant, degradable
Vinyl chloride	Resistant

micro-organisms present. Thus the presence of presticides or heavy metals, toxic for the biomass, will inhibit biodegradation of other organic compounds, even those readily biodegradable such as hydrocarbons.

On the other hand, certain substances reputed to be non-biodegradable under certain conditions, for example by aerobic or similar bacterial flora, may become biodegradable in some other conditions, for example anaerobic. This is true of certain chloro-solvents which cannot be directly degraded in aerobic conditions, but require a preliminary step of (partial) dechlorination of the molecules in the absence of oxygen. Once dechlorination has set in well, the carbon chains can be destroyed under the action of standard aerobic cultures. Through a detailed study on the degradation of trichloroethylene, Bourg et al. (1992) demonstrated the complexity of the concept of biodegradability in terms of experimental conditions; they also emphasised the complexity of the process of dechlorination, generating an ensemble of by-products from trichloroethylene, such as dichloroethylene, dichloroethane, vinyl chloride, chloroethane.... Under such conditions, it is necessary to consider in the feasibility study of the operation, the biodegradability of by-products or metabolites, which might form, and their toxicity vis-à-vis the environment. It would, of course, be futile to propose an operation of biodegradation of a complex toxic chemical to by-products which are as toxic as (sometimes more than) the original molecule, and non-biodegradable, or particularly stable over time.

In addition to the biodegradability of the contaminant, its concentration and distribution on the site are also important factors to be taken into consideration. In fact, when the quantity of contaminant concentrated at the same place exceeds a certain limit, the micro-organisms remain at the periphery of the contaminated mass and avoid access to it. Furthermore, for certain compounds a very large increase in the concentration can become toxic for the micro-organisms which are supposed to degrade them. One may thus have to dilute the contaminated material with a clean diluent to bring it within a range 'acceptable' to the micro-organism. Also, and especially for in-situ operations, the distribution of the pollutant on the site must be considered. Is it a matter of discrete sources of contamination, well defined in 'patches', localised but homogeneous within a patch for example? Or does the contamination correspond to an ensemble of poorly determined sources in which the concentration is heterogeneous and varies chaotically over a large area?

• The second factor to be noted is the **type of micro-organisms**. These may be bacteria, fungi, algae...; in most cases one has to deal with bacterial stumps.

For degradation of petroleum hydrocarbons, a large number of species naturally present in the soil, such as *arthrobacter, achromobacter, novocardia, pseudomonas, flabobacterium*—may be found useful. For the bioleaching of heavy metals, stumps of *thiobacillus, leptospirillum*, or *sulfolobus* are commonly employed. These stumps, resistant to concentration of metals, 'work' in very acidic conditions, by oxidation of ferrous iron and sulphur, thus liberating metallic ions, which are rinsed out of the medium.

During the study of feasibility of implementation of a biodegradation, the flora present in the contaminated soil is tested for its aptitude for degrading the pollutant present. This test is conducted in a laboratory, generally on a sample representing a few tens of kg of contaminated earth lifted from the site. To improve the efficiency of native flora, it is sometimes useful to grow such cultures in the laboratory from samples of natural soil for subsequent injection directly into the contaminated subsoil at the start of an in-situ operation. This is termed 'biostimulation'.

If the local flora is not adaptable to degrading the pollutant present (molecules whose degradation is difficult, e.g., PCBs, PAH) or is not very abundant, the technique to be employed consists of mixing into the contaminated material active but alien bacterial stumps. These micro-organisms, directly 'engineered' for specifically degrading a molecule or a group of molecules, must be adapted to the conditions of the medium in which they are to be inoculated, a penalising condition indeed. To facilitate better adaptation, the stumps to be inoculated are grown on supports (zeolites, carbonates, composite material...) and mixed directly with the material to be treated (on-site operation) or with water to be injected in the soil (in-situ operation). To test the adaptation of an alien flora to the medium and its efficiency, laboratory experiments have to be undertaken. Thus, Pellet et al. (1993) showed the effectiveness of 'engineered' flora for degradation of a soil polluted with PCB with an initial concentration of more than 140 ppm. In six weeks 90% of the PCB had been degraded in two tests in which the flora was introduced, while in the untreated (control) sample, the concentration of contaminant remained unaltered.

• **Choice of the electron acceptor and nutrients** always represents a key factor in the feasibility of a bioremediation operation.

A rich technical literature exists concerning the acceptors of electrons but to date opinion is far from unanimous. The term 'acceptor of electrons' refers to the capacity of the substances to accept electrons in the course of redox reactions used by the micro-organisms to transform organic compounds into energy. Under aerobic conditions, the acceptor of electrons is oxygen and the term 'aerobic respiration' is used in accordance with the equation:

$$O_2 \text{ (gas)} + 4H^+ + 4e^- \Rightarrow 2H_2O.$$

Oxygen may be supplied in several forms, say:
— atmospheric air,
— pure oxygen,
— ozone (O_3),
— hydrogen peroxide (H_2O_2).

Use of atmospheric air, added by diffusion or injected in water to be treated, is a priori not very effective, the quantity of oxygen necessary for achieving redox seldom being attained. Used in early experiments on biodegradation, it has now been practically abandoned. In comparison, the procedure called 'bioventing', which consists of injecting air at a small flow rate directly into the soil, appears very favourable, especially given the simplicity of application and its very low cost (the method is detailed later in this book).

Procedures using ozone or pure oxygen for augmentation of the level of oxygen dissolved in water are relatively simple; however, their application requires special technical conditions (generation of ozone, employment of liquid oxygen) which are not easy to maintain in the field; thus their application is restricted.

Hydrogen peroxide, on the other hand, has several noteworthy advantages over other forms of oxygen, whereby its use is favoured: unlimited solubility in water, ease of transport and putting to work, and facile utilisation by the micro-organisms. However, it does have two drawbacks: rapid decomposition in the subsoils and high cost. Its use these days is being curtailed.

Aside from oxygen, other acceptors of electrons can be used under conditions of low oxidation or when the process is not even actually aerobic. Thus in denitrification, nitrates are used as acceptors of electrons according to the equation:

$$2NO_3^- + 12H^+ + 10e^- \Rightarrow N_2 \text{ gas} + 6H_2O$$

The kinetics and growth of the biomas are, however, much less with nitrates compared to those with hydrogen peroxide, and this drastically reduces the efficacy of biodegradation achieved with this oxidising agent.

We may cite another two examples of employment of acceptors of electrons under anaerobic conditions: methanogenesis, fabrication of methane from CO_2,

$$CO_2 \text{ gas} + 8H^+ + 8e^- \Rightarrow CH_4 \text{ gas} + 2H_2O$$

or reduction of sulphates,

$$(SO_4)^{2-} + 9H^+ + 8e^- \Rightarrow HS^- + 4H_2O.$$

However, the redox potential in these two examples is much lower (of the order of 5 to 6 times) than that for aerobic respiration or denitrification.

For nutrients, the choice is less complicated, all investigators agreeing that the forms of phosphorus and nitrogen best assimilated by the micro-organisms are phosphates and ammonia. The major sources for nitrogen are thus ammonium salts (for example $(NH_4)_3 PO_4$) or urea ($CO(NH_2)_2$); nitrates are also used sometimes, for simultaneously exploiting their capabilities to act as the electron acceptors as well as nutrients.

• **Characteristics of the material or medium under treatment** are likewise to be considered in the feasibility of remediation by biological means.

In addition to the C/P/N stock already present and the quantity of oxygen available in the medium under treatment, other physicochemical characteristics can also influence the efficacy of biodegradation, namely, pH, temperature and humidity. In an on-site operation, these parameters are easily controlled and adjusted to maintain optimal conditions of application. But for an in-situ treatment they are difficult to control, let alone adjust as required.

Beyond certain critical thresholds, the activity of micro-organisms in a medium is slowed down (even arrested) if the conditions continue to deteriorate.

As for pH, it is advisable to maintain it in the range 5.5 to 8.5, albeit fungi by and large tolerate more drastic conditions than bacteria do. Sulphur-reducing bacteria, intervening at the stage of leaching/fixation of heavy metals, act at a much lower pH, of the order of 2 or 3.

The concentration of water in treated material should be held between 25 and 85%; its role comes into play at the stage of transport and biochemical exchanges. Should there be a tendency towards desiccation, the medium has to be rehumidified.

Finally, temperature can also exert a harmful influence on bacterial activity which slows down if the temperature falls significantly. In in-situ operations of bioremediation, the process is retarded in winter and even completely arrested if the temperature falls below 0°C. The range of maximum biological efficiency is thought to lie between 15 and 45°C.

For remediations carried out in situ, **certain physical parameters of the soil and aquifer** are determinative for the success of the operation. They comprise, among other factors, permeability of the subsoil, its homogeneity, depth of water table....

It is obvious that the most important characteristic is permeability; in fact, if the subsoil is not sufficiently permeable (presence of silt, clay) circulation of fluids containing nutrients and oxidising agent will be poor and oxygenation low, restricting the coverage and extent of biodegradation. It is generally thought that when the coefficient of permeability in the medium under treatment is less than 10^{-6} m/s, in-situ remediation by biological means will be difficult to carry out.

If the subsoil is heterogeneous (very dissimilar geological layers—gravel and clay; presence of lentils of sediments of low permeability...), the process will tend to be preferential, leaving intact pockets more resistant to remediation or the pollutant, totally untouched.

Similarly, if the water table is close to the surface (less than 2–3 metres), the thin unsaturated zone will tend to be less oxygenated as it is difficult to homogenise the spread of the additives delivered in the subsoil; contrarily, if the water table is deep (> 20 metres,) an effective oxygenation of the subsoil and a good percolation of the added substances becomes possible.

• **Conclusion: advantages and limitations of bioremediation**

In concluding this general section on bioremediation, let us summarise the main **advantages** of this technology:

— It is actually a technology of remediation, destroying the pollutants and not merely transferring them; the transformation generally generates CO_2 and water as final products.

— Treatment is complete, affecting all the media concerned (water, soil, soil-air) and various possible phases of the contaminants (liquid, solid and gaseous); furthermore, when remediation is carried out in situ, the entire subsoil is treated in one go (unsaturated soil, aquifer, groundwater).

— It is one of the more economical techniques, with the highest performance/price ratio (see chapter on costs).

— Its impact on public opinion is immediate.

However, it has some **limitations** also, which may be summarised as follows:

— The technique is applicable only to biodegradable substances and within the limits imposed by its implementation.

— When applied in situ, it is necessary that the subsoil permeability be higher than 10^{-6} m/s.

— If the pollution comprises a mix of several contaminants, biodegradation may be inhibited, some contaminants playing the role of toxic substances for the micro-organisms; in addition, transformation sometimes generates metabolites which are as—or even more—toxic and sometimes more stable than the original polluting compounds.

— The duration of application generally extends to several months for in-situ and on-site applications.

The bioreactor

The bioreactor represents a technique of remediation by biological means which can be used for numerous applications; the operation consists of biodegrading the contaminant in a container installed on site (closed box, basin, column...) by addition of the other ingredients necessary for the reaction. Through a bioreactor one can remediate:

— water pumped out beforehand,

— soil, treated in the form of slurry, or sludges,

— a gaseous phase (the term biofilter is used in this case).

Implementation of the technique remains the same for these three types of media; only the equipment may differ.

Two types of reactors are used in which:

1) The micro-organisms grow in suspended state in the contaminated environment and form a floc,

2) The micro-organisms grow in the reactor attached to a support.

In the suspended system, the contaminated water circulates in a basin in which a population of micro-organisms degrades the organic matter (comprising the pollutants) and grows by feeding on it. The floc formed is subsequently settled out and the biomass is recycled to the reactor.

In the system with support, the micro-organisms are attached on an inert material or structure provided or installed in the reactor.

For soil and water, this technique poses the same major drawbacks as do all other on-site treatments: water has to be pumped out or the soil excavated beforehand.

But the bioreactor has certain significant advantages:

— It enables precise control and management of the process of biodegradation; in fact, the systems of control of pH, humidity, concentration of nutrients...are easy.

— Mixing the material under treatment and micro-organisms on the one hand, and nutrients on the other, is readily and effectively accomplished; aeration of the ensemble is likewise generally easy.

— Optimal conditions of biodegradation can be quickly attained and hence high operational efficiency; reaction times can be continuously readjusted, depending on concentration of residual contaminant and metabolites and even the biomass present in the reactor.

— Depending on the pollutant and the material, the micro-organism best suited for the treatment can be determined (either part of the bacterial flora from the contaminated medium itself, or alien stump added in the reactor).

• **For the treatment of solid materials** (soil, dirtfills, wastes...) certain preparations are necessary, i.e., homogenisation, crushing lumps, sieving... (Fig. 5.29). This material is then mixed with water (generally 30% by weight/volume) and introduced in the reactor by pumping. Several modes of operation are technically possible—continuous flow or in 'batches'—depending on the situation. Regardless, bioreaction time has to be modu-

Photo no. 11: Bioreactors (source: BRGM).

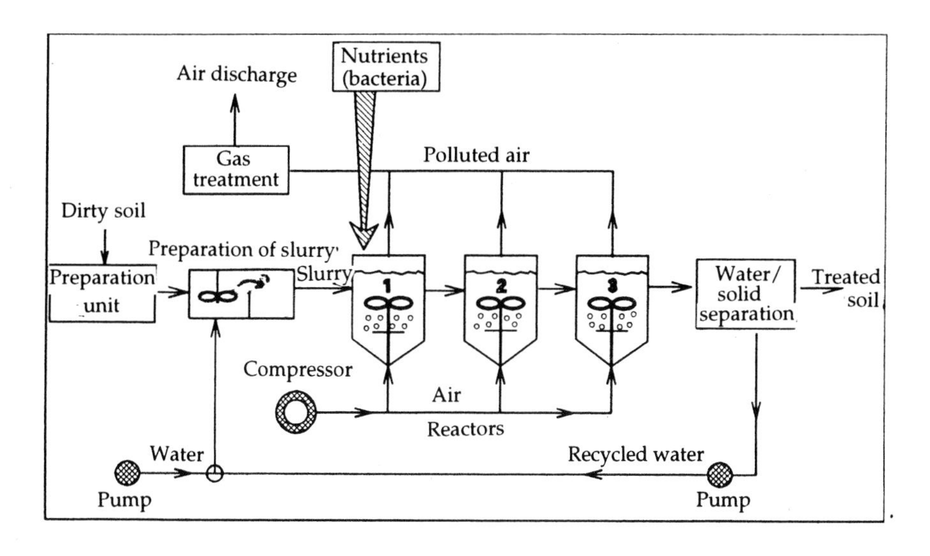

Fig. 5.29: Schematic sketch of bioreactor adapted to treatment of soils.

lated case by case, depending on type of contaminant present, its initial concentration and acceptable level consequent to treatment.

There are several reactors in a series in most apparatus, transiting the 'slurry phase' from one to the next. Nutrients are added in the first reactor in a quantity calculated in terms of the quantity of contaminant present, to obtain the optimal proportions of carbon, phosphorus and nitrogen. If necessary, micro-organisms are also added at this stage. Each reactor is provided with an agitator to ensure vigorous mixing of the ensemble of constituents of the reaction (rotating arm inside, rotation of the container...). Aeration, usually in the form of jets of air at the bottom of the reactor, completes the apparatus. The air, loaded with contaminating gaseous phase, is recovered at the top and directed towards a system of purification (biofilter for example) before its discharge into the atmosphere.

When slurry is treated, it passes through a solid/water separator; the solids are stored (later restored) while the water containing numerous adapted micro-organisms is generally recycled. This procedure reduces the time of acclimatisation of the micro-organisms and treatment for the follow-up operation.

This technique is particularly well suited for soils or wastes in which the concentration of organic contaminants is high, with a large proportion of an oil phase. This process is used, among other things, to remediate earth contaminated with PAH, phenols and, more often, petroleum products such as gas oil and fuel oil.

The efficiency of the reactor is illustrated by some results listed in Table 5.9 from Vogel (1993).

Table 5.9: Some results obtained by the bioreactor (extract from Vogel, 1993)

Compound	Initial concentration (in mg/kg)	Reduction (in %)	Time of treatment (in days)
Volatile organics	500	99	
Phenanthrene (PAH)	46	58	10
Total organic carbon	159,000	27	10

• **For the treatment of water** the principle is the same. Before its entry in the reactor, a readjustment of pH may be necessary (between 6 and 7.5) depending on the nature of the water as well as the addition of phosphorus and nitrogen. Two types of devices are in common use and are described below:

The first device is the 'in-column' technique. Water to be remediated is passed through a column (Fig. 5.30)—one or several in a series—in which it follows a zigzag path and comes into close contact with a neutral material. This material, constituted of polyethelene foam, or PVC, for example, serves as a support for the biomass. During this journey, the organic contaminants present in the water get biodegraded by the fixed micro-organisms. The growth of biofilms on the supports can be very fast and visible—an excellent measure of efficient biodegradation.

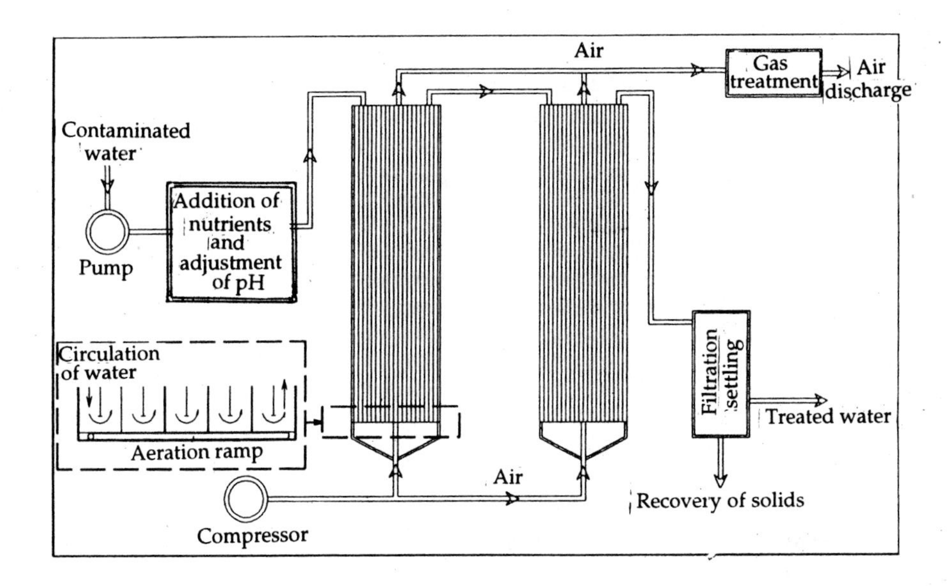

Fig. 5.30: Schematic representation of biodegradation of water by the in-column technique.

Should some undesirable metals be detected in the water, they must be removed before treatment is started. Precipitation of iron, if present in the solution, could even clog the system if not removed.

The result obtained depends greatly on the length of stay in the column, which is decided according to the quantity of contaminant to be destroyed. To modulate the time of stay, one can adjust the flow rate of the water injected (a few m^3/h generally) or the height of the column (a few metres) or both.

As in the case of solids, aeration is done through the bottom of the reactor and the pulsed air rises in the form of bubbles to the top of the column, eventually becoming loaded with a contaminated gaseous phase; in this case too the air is recovered and passed through a filter (biological or activated carbon) before being discharged into the atmosphere.

The other possible device is similar to the one used for soils; water is injected in a tank, open or closed, and is mixed with a pulverised support on which the biomass is affixed (zeolite, limestone, algae...); the mixture is constantly stirred during the time necessary for biodegradation of the contaminants present in the water. After the time required for reaction, the water is removed and separated from the biomass.

In both devices, the water is settled after treatment and/or filtered before discharge to the sewerage/drainage system or injection in the subsoil.

The results obtained are generally good. For example, for the biodegradation of pentachlorophenol (used for the treatment of wood from sawmills), more than 90% of the contaminant is usually destroyed in the treatment of both the soil and water (EPA, 1992) with an initial concentration in the soil of several hundred mg/kg; for PAHs the overall efficiency is slightly lower (of the order of 85 to 87% destruction).

• **For treatment of gas,** use of a biofilter increases the effectiveness and simplicity of the system. The biological filter is usually made of compost, an ideal medium for the multiplication of bacteria. Alternatively it could be made of peat. It can be installed in two alternative forms: the filter 'in bed' or the filter 'in packet'. The latter (Fig. 5.31) consists of a layer of humidified compost held between two grills in a closed tank; air loaded with pollutants is injected from below at low pressure, through a layer of coarse porosity (gravel, plastic stand), which facilitates its uniform distribution in the tank. After passing through the compost, air is recovered from the top of the tank. To complete the installation, a tap is provided at the bottom to recover the water of condensation and a sprinkler above the compost to humidify as per requirement.

The efficiency of the biofilter depends directly on the thickness of the compost filter and, of course, on the velocity of the injected gas. Easy to use, this type of biofilter is interchangeable and several tanks can be installed in series, depending on the concentration of contaminants in the inlet gas. Before passing through the filter, the gas is humidified, for example by passing through water (until relative humidity reaches 80 to 100%). If the filter is maintained in good condition, up to 90% of the contaminants present (especially BTX) can be degraded. The major factors for success of the compost bed are 50 to 70% humidity, porosity between 80 and 90%, pH 7 to 8 and temperature varying from 15 to 45°C.

Biodegradation in 'pile'

The term 'biodegradation in pile' covers several methods of biological remediation, some of which have been in use for a long time. We shall describe here the methods of composting, 'land farming' and the technique called 'biopile'—all expressions borrowed from the Anglo-Saxon technical literature.

Application of these methods is by definition accomplished on site; only solid materials, primarily soils contaminated by organic substances, are covered.

The basic process of biodegradation in pile consists of excavating the soil and bringing it to a nearby treatment unit designed to facilitate natural aerobic degradation. If acceleration of the process is desired, three influencing factors have to be taken into account and controlled irrespective of the technique used. These factors are aeration, level of humidity and supply of nutrients. The source of micro-organisms is usually the bacterial flora present in the soil but micro-organisms of other origin may also be added.

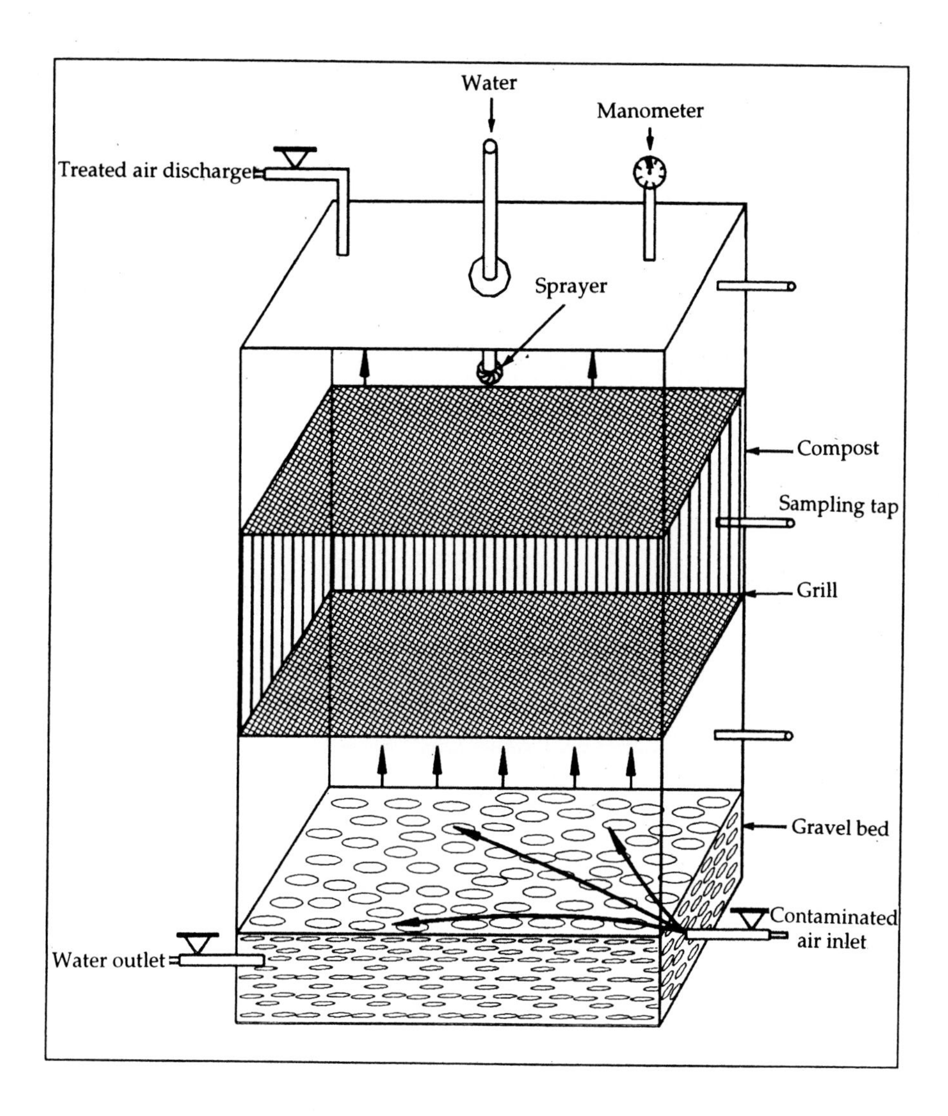

Fig. 5.31: Schematic sketch of a biofilter 'in packet'.

• The simplest technique is that by **composting.** The excavated soil is disposed in regularly spaced swathes (small heaps) a few metres in circumference and about one metre high (Fig. 5.32). In a humid temperate climate natural conditions suffice for humidification of the swathes. In these conditions, however, the process of degradation is very slow and not very cost-effective. To expedite the phenomenon, it is customary to mix earth with a coarse organic substratum which facilitates aeration in due course of time and supplies a nutritive complement to the reaction; one can use materials such as straw, bark pieces, undergrowth, manure.... Biodegradation of the organic matter introduced generates heat which enables biological activity to take place almost throughout the year. Excavation, sieving and piling also facilitate the aeration of the material for starting the process.

The efficiency obtained is generally low; the technique should rather be used only for easily biodegradable contaminants (petroleum-based hydrocarbons, for example).

When the compost produced is discharged, the volatilisation of some pollutants and their release into the atmosphere may create a negative impact on the environment. Also, this type of technique can be applied only on pollutions composed of slightly volatile substances (gas oil for example).

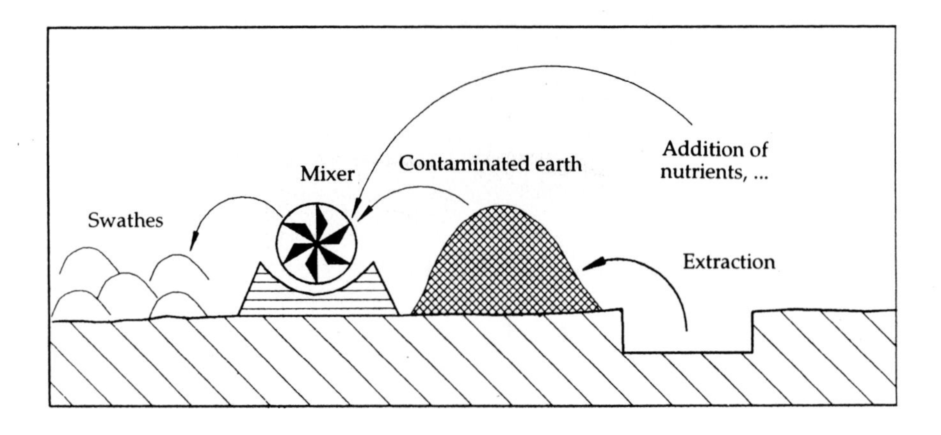

Fig. 5.32: Schematic representation of composting.

• A more elaborate technique is that of **land farming** which consists, as its name implies, of treating the material as an agricultural soil to facilitate its remediation. This method, a bit more complicated than composting, requires the following operations (Fig. 5.33).

— The contaminated soil is first spread over a large plane surface to a thickness of a few tens of cm, making it possible to subsequently knead it with agricultural tools; to preclude the risk of recontamination of the adjacent lower soil layer, impermeable surfaces are selected for spreading the contaminated soil; in suburban and industrial zones asphalted parking lots have proven highly suitable for this purpose.

— A fertiliser is then added, either chemical fertilisers or organic manure; it is spread over the entire surface to be treated and is subsequently kneaded into the contaminated soil. The addition of fertiliser improves the balance between the nutrients and the source of carbon and, in

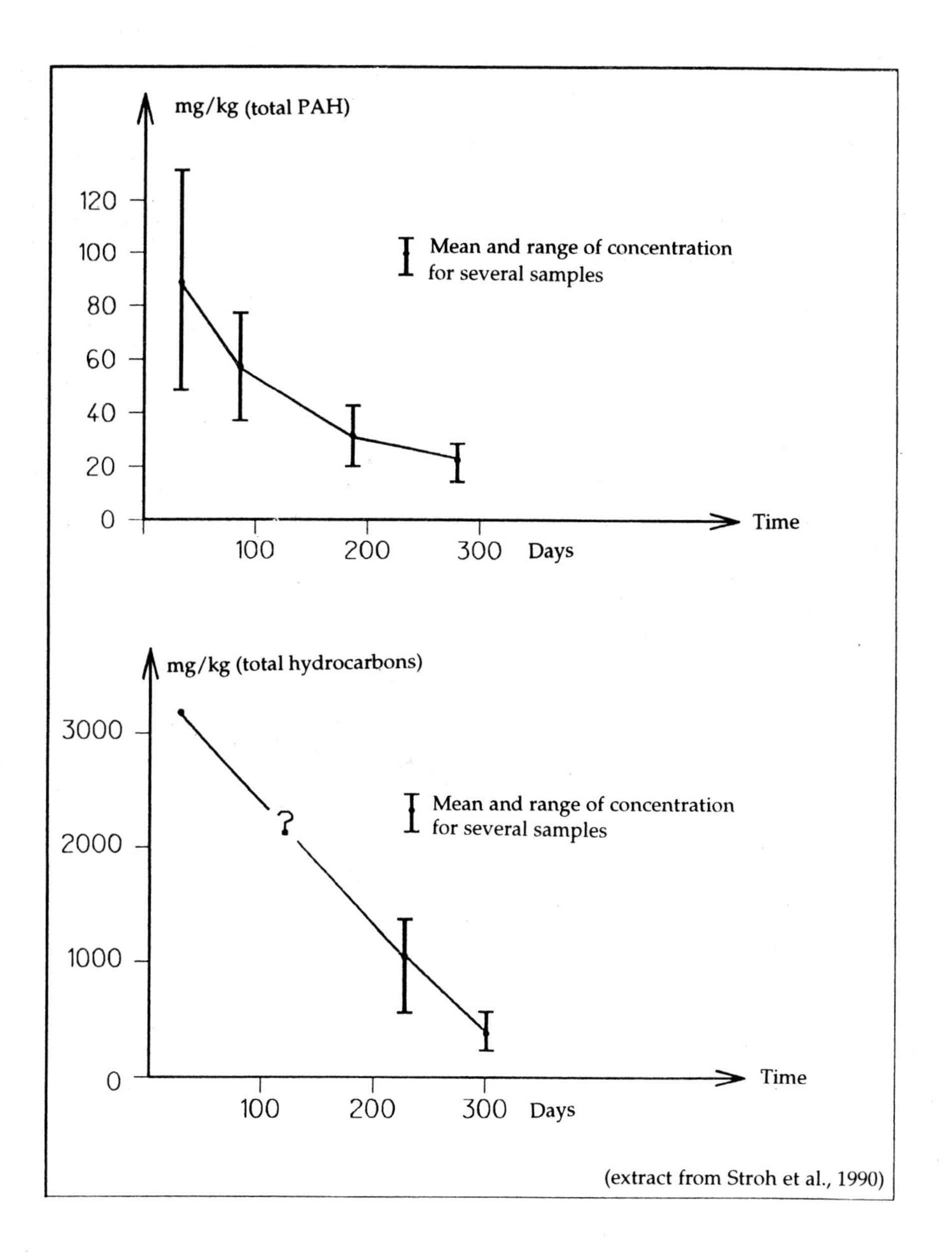

Fig. 5.35: Graphic illustrations of efficiency of biopile (reproduced with permission of Kluwer Academic Publishers).

in situ, neither excavation of the soil nor pumping of water is required. Compared to on-site methods, the biotechnology applied in situ has two main advantages:

— on the one hand, the subsoil and groundwater are treated simultaneously,

— on the other, the costs involved are lower (see relevant chapter).

The in-situ biological treatment is preferable in some cases, for example when it is necessary to remediate underneath a building (Fig. 5.36a), or when the contamination is located at considerable depth (say, several tens of metres from the surface—Fig. 5.36b). Another typical case when in-situ treatment is preferable is that of lateral dispersion of a contamination by petroleum hydrocarbons. In this last type of contamination, the substances percolate down by gravity to the surface of the water table and then disperse laterally over a rather large distance (Fig. 5.36c).

Further, compared to on-site treatments by biodegradation, in-situ bioremediation requires perfect expertise in the biological process which develops in the subsoil. Sound knowledge and comprehension are imperative not only of the action of the bacterial flora in the soil, but also of the dynamics of the hydrogeological system.

The **feasibility of an in-situ treatment** thus constitutes a complex issue and each situation should be considered individually in terms of its specific characteristics. For this reason, it is often difficult to determine a priori whether in a given situation, in-situ biodegradation would be a cost-effective alternative. Also, in this domain 'ready-made recipes' applicable to all situations, do not exist.

We have already discussed in the earlier subsection dealing with feasibility of biodegradation the parameters that must be taken into consideration while deciding the applicability of the technology. Without repeating the details of these parameters, let us emphasise once again the importance of a sound knowledge of the characteristics of the subsoil; for an in-situ biotreatment these characteristics are decisive for the success of the operation. Computational models exist whereby the viability of an in-situ biological treatment can be estimated. These models take into consideration the relative influence of all the parameters involved. A score is assigned to each parameter and the sum of such scores is checked against a relative scale which enables determination of whether or not the treatment would be practical. Such a model has been developed in particular by the American firm Remediaton Technologies Inc. of North Carolina.

In this particular case (Anon., *The Hazardous Waste Consultant*, January/February, 1992), the specific characteristics of the subsoil are determinative on the one hand, at the hydrogeological level (permeability, heterogeneity and thickness of the aquifer...) and on the other, at the level of chemistry of the subsoil (pH, presence of certain anions, heavy metals...). Their importance is assessed by the scores attributed to each as per the

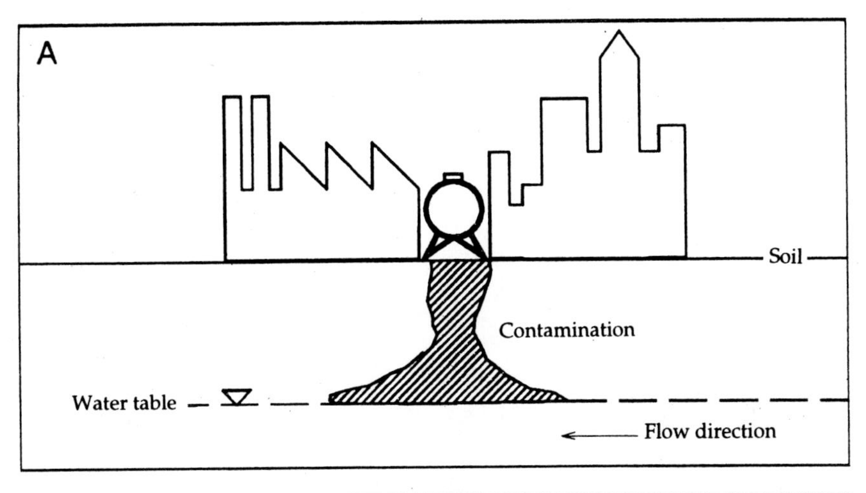

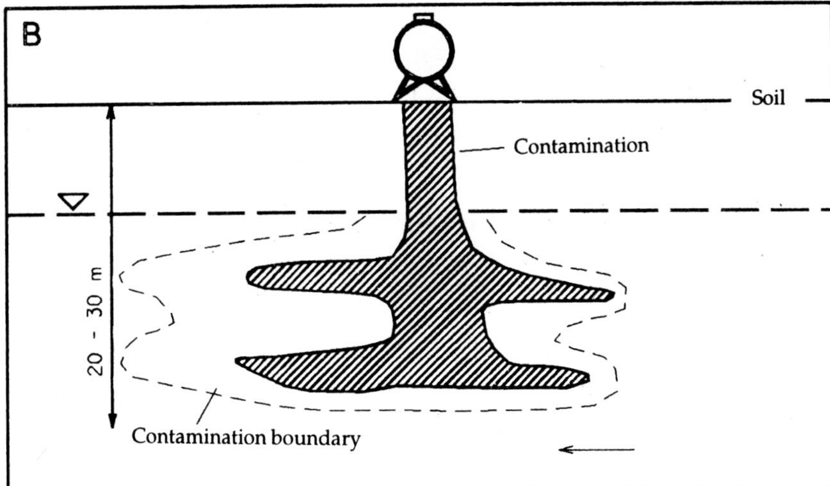

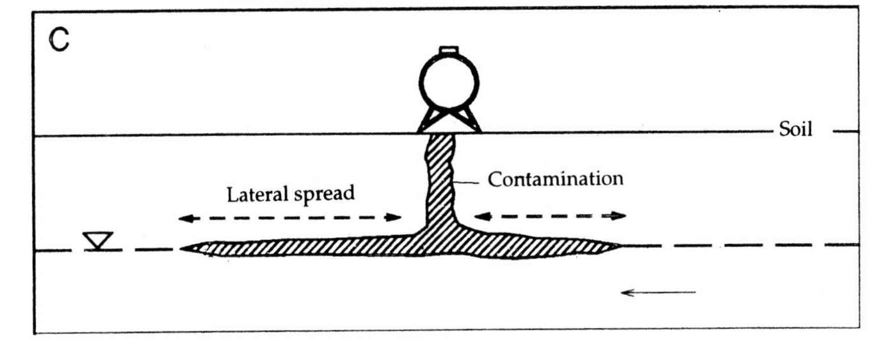

Fig. 5.36: Dispersion of a floating organic contaminant—situations favourable for in-situ bioremediation.

Photo no. 13: Unit of biodegradation for in-situ treatment (source: BRGM).

general list: thus hydrogeology and chemistry cumulate relative scores of six while the characteristics of the contaminant (type of substance, source of contamination, concentration present...) receive relative scores of only two.

• **The procedure applied** in an in-situ biological treatment consists of introducing in the contaminated zone the requisite amounts of nutrients and acceptors of electrons; the commonly followed method is to inject water in the subsoil in which phosphorus, nitrogen and oxygen are dissolved, the aerobic reaction being the one most widely used in this case. The doses of additives are calculated beforehand in terms of the source of carbon present and the flow rate and injection times expected.

Three schemes of utilisation may be possible (Fig. 5.37):

i) if the contamination is located in an unsaturated zone,

— the solution containing nutrients and oxygen is infiltrated in the soil from the surface (spraying, lagooning...), provoking, by its passage, growth of the biomass and degradation of the contamination:

ii) if the contamination is located in a zone at the level of the groundwater or in the groundwater,

— the solution is injected in the groundwater (at its surface if the contaminant floats on the surface of the water table), upstream of the zone under treatment; it then flows downstream by the groundwater flow and runs through the contaminated zone being treated by it—this is a passive system;

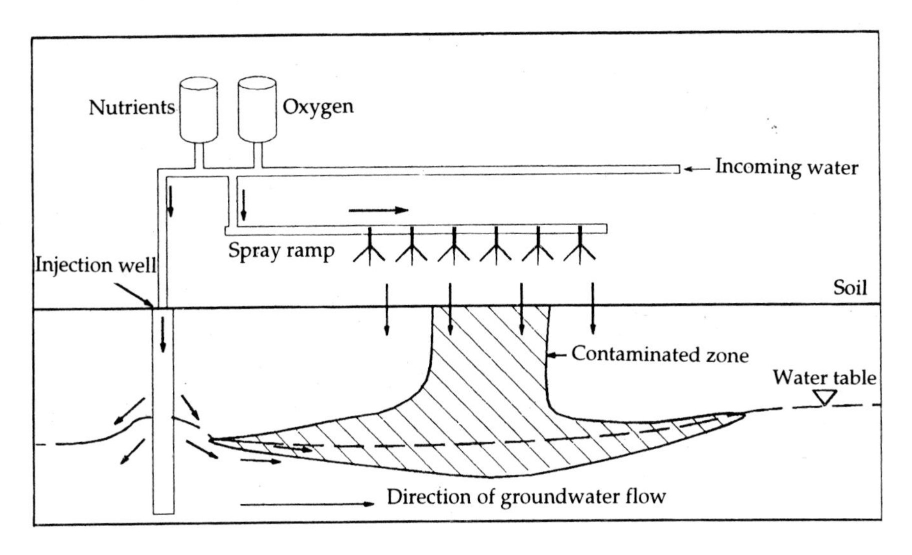

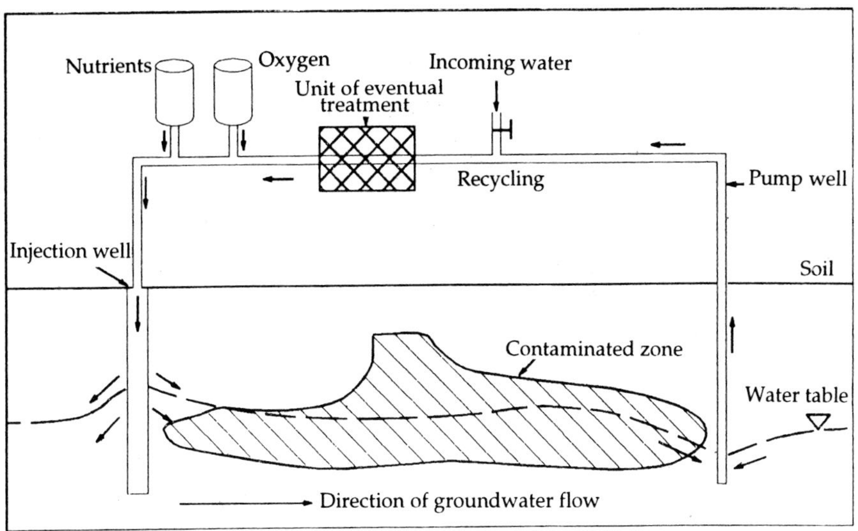

Fig. 5.37: Schemes of implementation of in-situ biodegradation.

— the solution is injected upstream as in the preceding case, but also farther downstream of the zone, water is pumped out and reintroduced in the circuit at the upstream injection point—this is an active system.

The latter system is more efficient because a motion by convection is generated, activating circulation of water in the subsoil, while in the passive system, only natural motion is depended upon, hoping that it will suffice to carry the additives for treatment to the entire zone and thereby activate the bacterial flora.

In most cases the native flora are adequate. Sometimes, to accelerate the start of the process, it becomes necessary to stimulate their growth in the laboratory from samples taken and then reinoculate them in the medium (biostimulation). Sometimes, too, alien stumps are inoculated in the medium, knowing full well that their adaptation to the in-situ conditions is a priori not easy (bioaugmentation).

The design of the system often requires modelling by computer, especially to simulate the conditions of injection, flow and pumping of solutions. Employment of such modelling facilitates taking into consideration (and varying them as necessary) the basic parameters, thus ensuring the efficacy of the system. This is particularly important in determining the number of injection wells, their depth, spacing...and the system-design and operational plan to cover the entire zone under treatment.

BRGM has conducted operations of in-situ biodegradation by developing two types of procedures, active and passive.

Photo no. 14: Line of injection in an operation of in-situ biodegradation (source: BRGM).

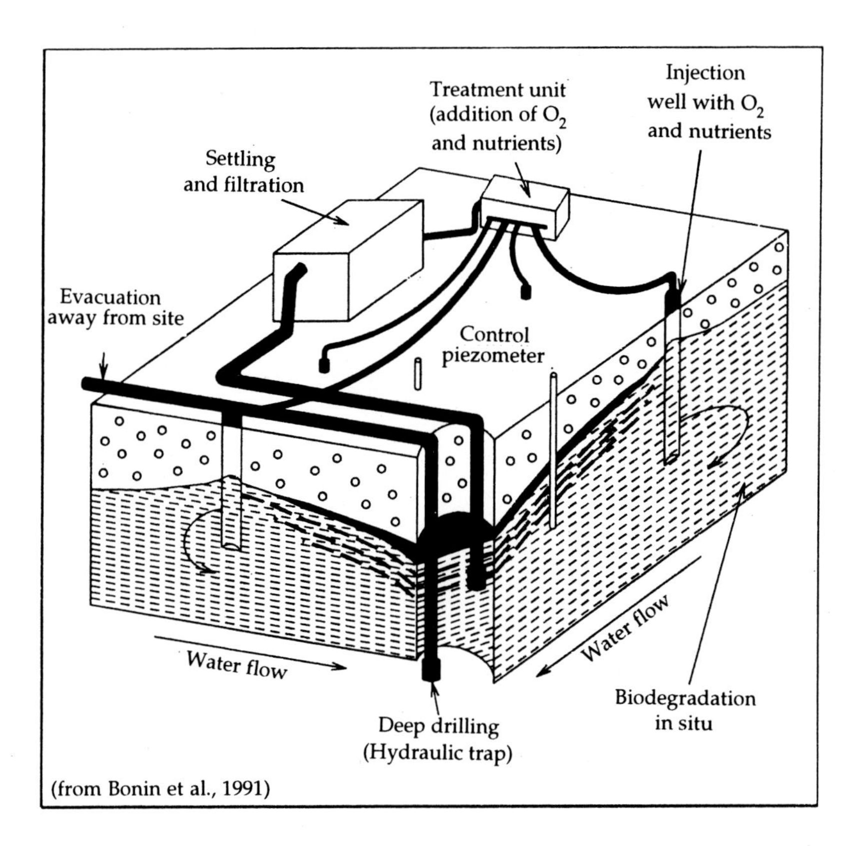

Fig. 5.38: Port aux Pétroles—schematic plan of the system employed.

At Port aux Pétroles of Strasbourg (Bonin et al., 1991), a pilot plant for in-situ biodegradation was installed in a zone contaminated with 65 m^3 of domestic gas oil accidentally spilled in the subsoil. After pumping about 50% of the quantity of hydrocarbons, the pilot plant was put in operation in an 'active system', with a central pumping well and 8 wells for injection distributed uniformly at 10 m around the central well (Fig. 5.38). The pumped water-oil mixture was separated at the surface and nutrients (phosphate and nitrate) and oxygen were added to the water. This water was then reinjected in the 8 peripheral wells. The hydraulic circuit created thereby aided good circulation of the solutions injected in the contaminated zone at the surface of the groundwater, at a depth of about 3 m.

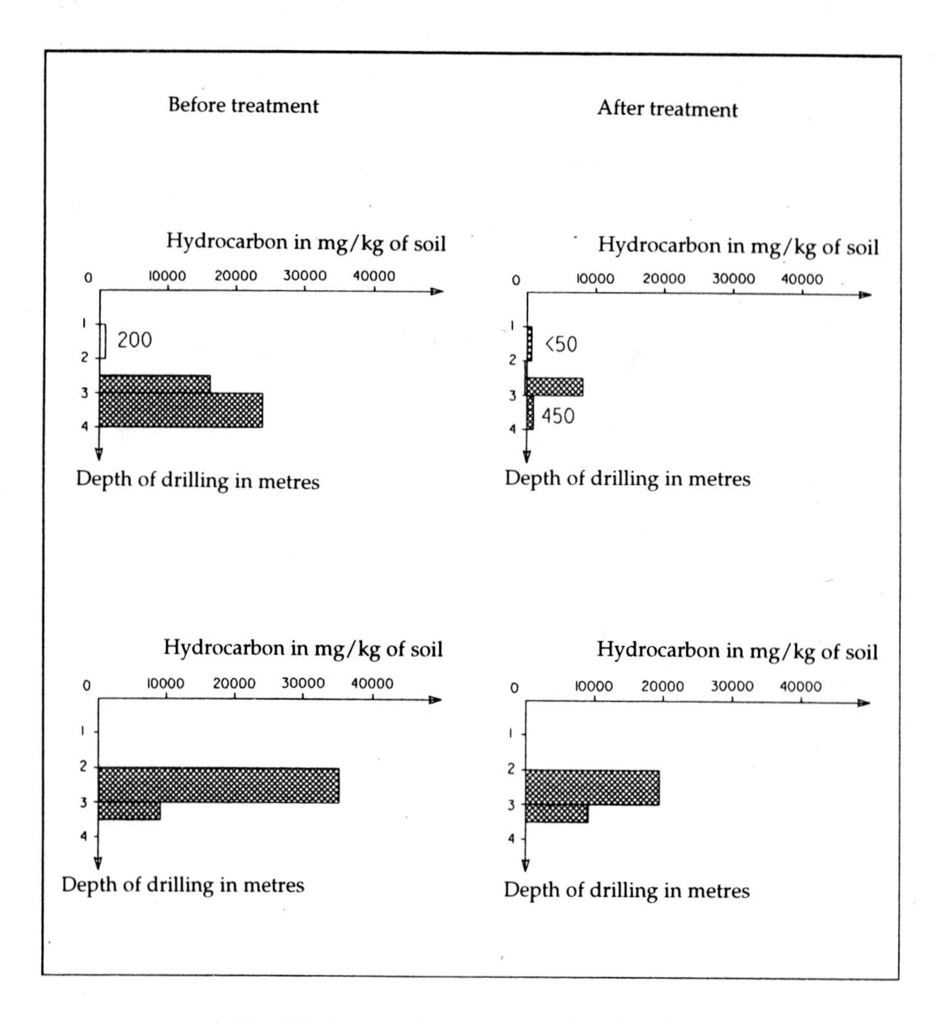

Fig. 5.39: Port aux Pétroles—illustration of results.

The experiment lasted 4 months, at the end of which a reduction of the order of 50% in the concentration of hydrocarbons in the soil was observed. The initial concentration in the soil was of the order of 15 to 20,000 mg/kg; at the end of the experiment, it varied from 7 to 9000 mg/kg (Fig. 5.39). In addition, the chromatograms showed complete disappearance of peaks of alkanes up to the 22 carbon homologues in all the samples analysed. Simultaneously, augmentation of the order 100 to 1000 times more in bacterial population and emission of CO_2 (increasing from 4 to 15% in the soil atmosphere) was observed.

On another site contaminated by petroleum hydrocarbons (fuels, petrol) remediation by in-situ biological means was carried out on a zone several tens of metres long and about 50 metres wide.

The contamination was characterised by concentrations of total hydrocarbons of about 2 to 10 g/kg in the soil; the values observed on the surface of the groundwater corresponded to a concentration in water of the order of a few hundred mg/litre. The groundwater lay at a depth of just a few metres from the soil surface. Geologically, the subsoil consisted of sandy silt surmounting a layer of coarse gravel; the thickness of these layers varied widely from one location to another in the zone, which influenced the water flow and thus the efficacy of biodegradation.

The system used was of the 'passive' type; a series of short drains was installed in tandem, perpendicular to the general direction of groundwater flow and upstream of the zone under treatment. The drains, uniformly spaced, were dug to the level of the water table. Uncontaminated water was injected in each drain at the rate of a few cubic metres per hour. Nutrients and oxygen were added to the water continuously in the requisite doses before its injection.

A number of sampling piezometers for control were installed to cover the entire site; this made it possible to supervise the progress of remediation by checking the numerous parameters in the groundwater, such as pH, temperature, dissolved oxygen, concentration of nitrogen and phosphorus, quanta of hydrocarbons dissolved in the water or emulsion on the surface of the groundwater....

In addition to supervising the quality of the water, carroting of the soil was regularly carried out to follow the biomass growth and change in concentration of hydrocarbons in the terrain.

After some months of implementation, a significant reduction, of the order of 65 to 95%, was seen in the concentration of hydrocarbons around most of the injection drains.

'Bioventing' and 'biosparging'

These two terms are drawn from the North American literature and pertain to techniques which combine the two mechanisms of remediation: biodegradation and ventilation.

By bioventing, we imply a forced aeration in the unsaturated soil (situated above the water table); on the other hand, biosparging implies direct injection of air in the water table (Fig. 5.40).

The supply of oxygen aids growth of the biomass, as happens in the case of a standard bioremediation, by inducing consumption of the carbon of the organic compounds present in the soil and consequently their degradation. Supply of nutrients is generally necessary to balance the

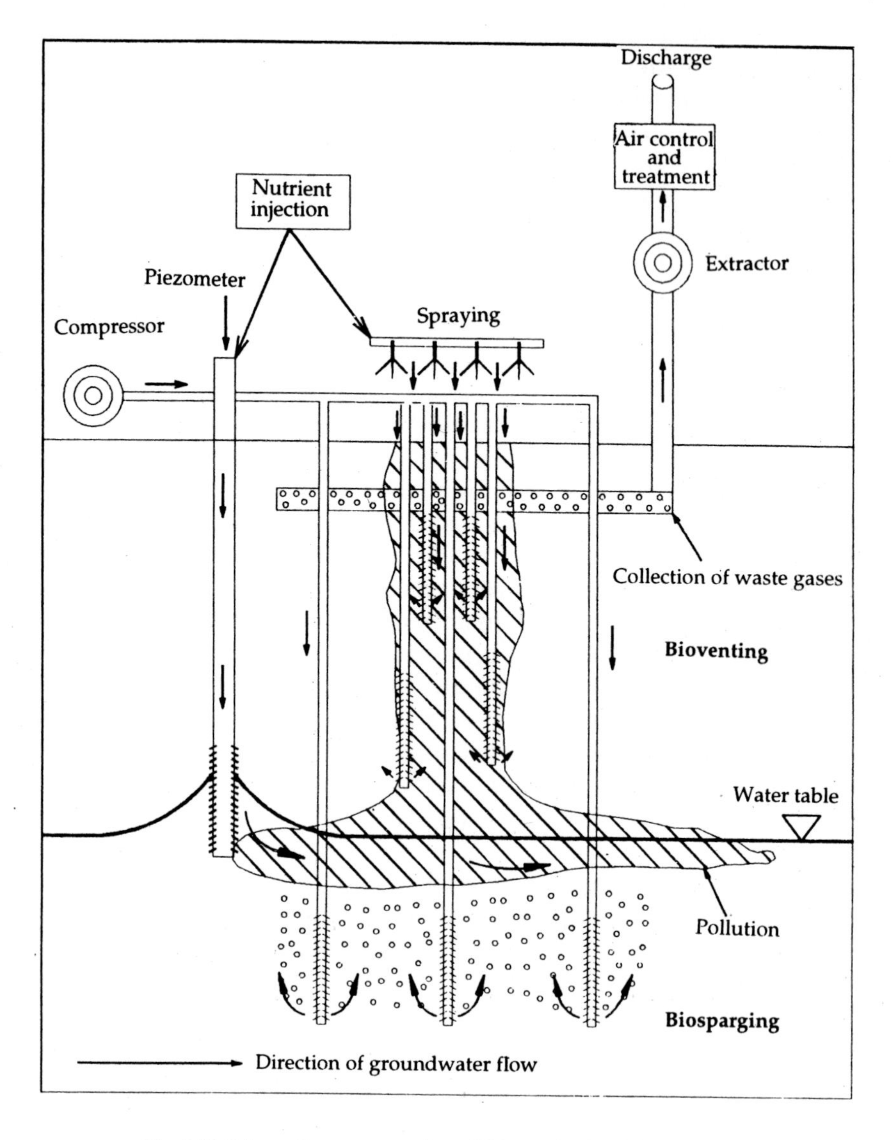

Fig. 5.40: Schematic representation of 'bioventing' and 'biosparging'.

proportions of carbon, nitrogen and phosphorus. These, mixed with water, are either injected in the groundwater upstream of the zone under treatment, or infiltrated directly in the soil of the zone by spraying or lagooning.

In bioventing, the volatile compounds of the soil are mobilised by the air current and this proceeds simultaneously with biodegradation. In addi-

Photo No. 15: Pilot of biosparging (source: P. Lecomte)

tion, as the gaseous compounds degrade, the injection of air aids volatilisation of the liquid phase.

In biosparging, the injected air mobilises the contaminants dissolved in water or trapped by capillary action in the pores, by vaporising them. This vaporised phase is carried upwards by the rising air bubbles and is biologically degraded in the unsaturated soil.

The airflow injected in the subsoil is recovered by suction. The flow rate of injection/suction is maintained sufficiently low to provide the bacterial flora of the soil sufficient time to degrade the contaminating compounds volatilised and entrained by the air current. This technique thus helps suppress (or reduce) the need for treating the gaseous effluents extracted from the soil, before discharge into the atmosphere. As for the technical design, this implies that only a small device of ventilation/suction need be used, instead of the bulky pulsators and air extractors usually employed in techniques of venting or evacuation.

The major difficulty lies in the follow-up and control of the process, access to which is only indirect and punctilious. The low permeability of soils as well as the presence of layers or strata of clay between more permeable zones, is a constraining factor. The presence of such layers can cause lateral dispersion of pollutants in the unsaturated zones or aquifers depending on the situation and thus extend the polluted zone. Besides, injection of compressed air in the subsoil can lead to formation of preferential channels of circulation (particularly true in the case of biosparging). If they are wide and therefore small in number or if they are not uniformly distributed in the volume to be treated, the efficiency of the process is drastically reduced. A bad structuring of soil and its lack of homogeneity also constitute major constraints for the use of techniques of bioventing or biosparging.

This technique is currently being employed more often, particularly to remediate soil contaminated by petroleum hydrocarbons. Excellent results have been obtained, with higher than 90% rate of degradation in a few months (sometimes even without the addition of nutrients). As for concentrations in the subsoil, reduction by several thousand mg of hydrocarbons per kg of soil is commonly achieved. In the USA, for example, bioventing is now generally used to remediate in particular fields of military aviation or zones of army fuel depots.

Biobarriers and biological screens

This is a recently developed technology and its aim is the in-situ treatment of contaminated groundwater. It consists of creating downstream of the pollution (and on the route of travel of the groundwater) a zone rich in micro-organisms well suited for removing the contamination to be treated. This zone of intense bacterial activity constitutes a real biological barrier, acting as a screen against the propagation of the pollution because on the passage of the groundwater, the contamination dissolved in water is degraded by the biomass which the flow of water must cross. Very often it primarily involves only bacterial microflora, indigenous in the soil of that area.

Many types of devices can be considered to serve this purpose.

In one such type, the biological barrier is installed directly in the matrix of the aquifer whose permeability, if it is not sufficiently high, is artificially increased by fracturisation to allow the creation of ecological niches. The increase in biomass is controlled by injection at regular intervals (for example through closely spaced holes drilled in a line) of a nutritive solution (also including an acceptor of electrons and/or a carbon source of energy). Another system consists of placing a ramp of biosparging (see the preceding section), with injection through drains or crevices. Also, the polluted zone may be either isolated at its upstream end by a physical barrier, or the plume of pollution may be physically channellised towards the biological barrier. A certain number of critical parameters of the effective biodegradation of pollutants, such as the volume and dilution of injected nutritive solutions, the duration of and interval between injections etc., should be determined separately in each case.

This technology is used for waste water or water pumped out from mines or water from drains in zones of storage of mine wastes (slag heaps, lagoons). The principle consists of developing downstream of these zones, marshy stretches into which the effluents flow. In the course of crossing this swamp, the heavy metals, present in large quantity in the effluents, are immobilised by the action of the flora present (bacteria, algae, reeds and macrophytes rising above the marsh...) and acidity of the effluents is reduced. Applied to waste water, the action of micro-organisms and vegetation provokes degradation of organic matter and denitrification of the water. This type of treatment, administered to urban sewage, is called the technique of lagooning and has been tried for effluents generated by wood-treatment factories also.

Treated effluents of acceptable quality are obtained which may be discharged to the drainage network, without risk of poisoning the natural milieu.

Water from mines is usually very acidic and heavily loaded with metals. In the course of natural oxidation of sulphides present in the ores, heavy metals get ionised and dissolved in mine water and sulphur gets converted to sulphuric acid.

While implementing the technique of humid zones, a basin temporarily holding the effluents and rich in organic matter and vegetation is created, wherein anaerobic conditions develop (Fig. 5.42). Such conditions favour reduction of compounds earlier rendered soluble by oxidation. This particularly induces precipitation of metals in the form of sulphides, due mainly to the catalytic action of bacteria (also called sulphur-reducers). Metals immobilised in the form of sulphides may be allowed to remain in this form without prejudice to the medium, as long as appropriate conditions are maintained.

The phenomenon also induces neutralisation of acidity of water, notably by the large production of gaseous H_2S, related to bacterial activity. A large augmentation of pH, exceeding 2 to 5 or even more, may be observed downstream.

Although bacterial sulphur-reduction is the dominant mechanism, the action of other groups of vegetation is not negligible; for example, absorption of metallic ions by algae or higher plants is quite effective; filtration of suspended particulate matter is carried out even in the tangle of roots or immersed parts of the vegetal cover. In this context, compost, woody undergrowth and debris, fungal mycelia and, above all, peat, significantly expedite the efficacy of the system.

The design of such a system has to take into consideration numerous parameters, such as extent and areal size of the device, its depth, the most appropriate associated vegetation to be introduced.... These parameters themselves are highly dependent on the quality of effluents under treatment and the level of concentrations of pollutants encountered.

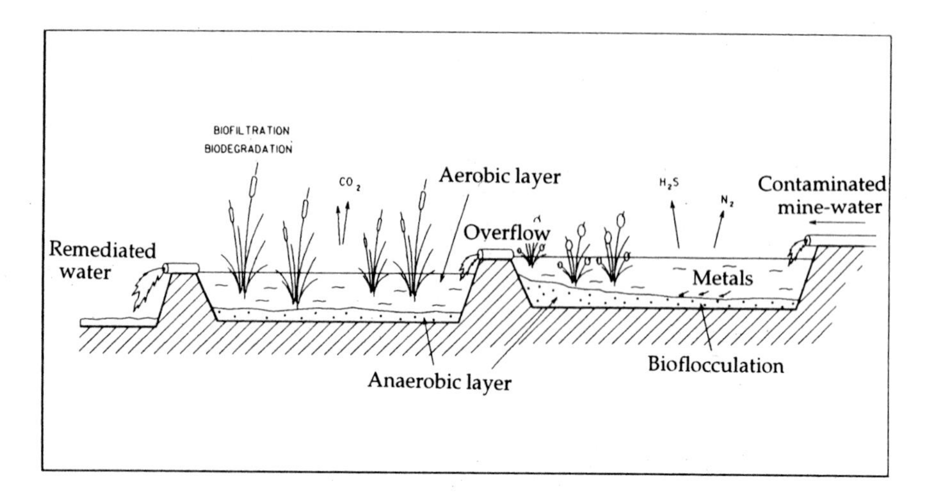

Fig. 5.42: Explanatory scheme of the biological method of 'humid zones'.

Numerous applications of this technique have shown that the physico-chemical quality of the water at the downstream exit of 'constructed' marshes (in other words at the 'overflow level') is acceptable in many instances for its release to the natural medium.

Thus, for example in France, the urban sewage flowing into the pond of Tau (Sète) first passes through a series of lagoons which effects its progressive treatment and upgradation, before reaching the last pond which remains uncontaminated.

In the USA, this method is commonly used for treating water from coal mines in the mining district of Tennessee, where marshy zones have been constructed downstream of the mining sites. The lagoons are large in area but small in depth (half a metre) and arranged in steps. A vegetation, becoming denser and more varied, covers the zones from upstream to downstream and expedites immobilisation of large quantities of iron, manganese and sulphur present in the raw mine-drainage waters.

This type of treatment is passive and hence does not require much attention; once installed, it suffices to let the vegetal mass which develops in the lagoon act. This is a reasonably efficient processs, particularly well suited to immobilisation of heavy metals and reduction of acidity caused by effluents from mines. However, like all treatments dependent on biological growth, it can weaken over time. This is particularly the case if its capacity is overestimated in relation to the quantity of metals to be immobilised. It has also to be borne in mind that the technique requires a large surface area for effective functioning.

VI

Costs

Even though the 'cost' consideration is not the main deciding factor in the planning of an operation of remediation, it is at least one of the determinant aspects which, very often and at the final stage, tilts the balance in one direction or the other, when choosing a plan or an action of remediation.

Numerous attempts to synthesise and compare the cost of various techniques of remediation have been made, more particularly abroad, in the USA and northern Europe (Holland, Germany...).

A comparison is generally difficult, if at all possible, because it involves comparison of factors which are not necessarily comparable. For example, how does one compare (i) a quantity of earth excavated and the solvent 'bath' for extracting remediated from it, with (ii) a plot of terrain which is remediated by evaporating the volatile compounds, or with (iii) a refuse dump which is confined all along its periphery by erecting a moulded wall? In practice, it is not an easy task to determine whether one technique is more complex than another and, depending on the criteria selected for comparison, one eventually obtains different results.

The choice of a criterion common to all methods would constitute, as a matter of fact, the key to success of this type of comparison. The most commonly used indicator consists of expressing the cost of treatment per unit volume (or mass) of the material treated. This reduces the comparison of various techniques to consideration of the cost of remediating a ton of earth excavated or a cubic metre of water pumped.

Once the issue of the criterion for comparison is settled, another major difficulty is in deciding which factors will be considered for calculating the cost. Depending on the factors taken into account, the result may differ significantly. Usually the following factors are not included in this calculation:

— Study of characterisation, including diagnosis of the site, analysis of risks and feasibility study of the plan of remediation. However, if prefeasibility tests need to be conducted on a large scale, their cost may be included.

— Follow-up action and control of operations of remediation by an independent expert agency, which will ensure good management of the project and be responsible for drawing up the final balance-sheet.

— Restoration of the site after the operations are complete, comprising, among other things, filling in excavations, restoring the approach to the site, greening and so forth.

On the other hand, the cost of remediation includes all factors associated with the treatment itself, and in particular:

— installations and transport of materials (expenses of transport, commissioning of machines...);

— consumables, for example activated coal, geomembranes, nutrients, solvents...;

— ensemble of fluids and energy: electricity, water, steam...;

— employment of technical personnel responsible for carrying out the operation plus expenses of premises needed for the operators;

— write-off of material, equipment, installations...; in some cases this is identified as a lease-purchase, the material becoming the property of the client after an agreed period of use in treatment.

One may also include in the unit cost, the expenses of maintenance of installations for an agreed period, for example in the case when after a first treatment, the system of remediation has to be kept operational.

Another difficulty pertains to extrapolation of costs from one country to another; for example, could a cost calculated in the USA be directly converted for application in France? and if not, what factors must be taken into account for such conversion?

As a matter of fact, differences in the following contexts:

— economic (state of national markets and competition among operators for example);

— legal (variation in regulations regarding application of certain technologies);

— geographic (consideration of much larger distances in one country compared to another);

— climatic (wintery conditions during long frost period);

— etc.

can lead to variations in behaviour and reaction of all the parties concerned, which influence the cost of remediation of sites markedly, rendering comparison between two countries rather difficult.

Thus, in Great Britain a refuse dump for industrial wastes is considered a relatively easy solution because obtaining permission to install a dump on site, under very favourable economic terms, poses no problem. In France, on the other hand, installation of a refuse dump is considered one of the very expensive solutions. Centres for Technical Burial of Class 1 (reserved for toxic industrial wastes) are few in number, very strictly regulated and managed by specialised companies. Current legislation is directed towards maximum reduction of such dumps, to encourage limitation and reclamation or destruction of wastes to the greatest extent possible.

Finally, estimating the time needed to remediate a unit volume is difficult. As a matter of fact, for many methods the time factor is fundamental to estimating the cost; thus the techniques employed 'in situ', such as venting or biodegradation, necessitate taking into account the concentration of the contaminant, subsoil characteristics and so forth. For some methods, contrarily, the time factor is not important, the technology being applied once and for all: confinement, refuse dump, incineration etc.

All the aforesaid difficulties prompt prudence in estimating the costs discussed here and allowing a large margin in prices. On the other hand, instead of considering just the estimates cited in the literature, data based on experience in France have been compiled and compared with numerous references existing elsewhere; thus our own scale of values was evolved. Table 6.1 summarises the range of costs for the most commonly used techniques classified according to the scheme developed in Chapter 5.

Between 1995 and 1998, the overall costs have not changed much, the factors leading to rise in costs and those leading to fall have mutually compensated. Two phenomena are, however, worth taking note of: on the one hand the relative stagnation of markets tends to keep the costs unchanged and on the other, the foreign firms, already well established in their country and wishing to start operation in France, tend to keep the costs low—though the same is not the case with the national firms dealing with remediation who would wish returns to go up.

One can see from Table 6.1 that the techniques of 'evacuation' (venting, pumping etc.) or biodegradation are, generally speaking, less costly than those in the other groups. It can further be seen that the Table presents relatively cheaper methods: enervation or confinement, chemical washing without expensive additives, or even application of on-site thermal treatment such as desorption for example.

It may also be noted—as already pointed out at the beginning of Chapter 5—that the methods applied 'in situ' are usually less expensive than those applied on site as well as those in which treatment is done away from the site in specialised units (special dumping, incineration etc.).

Passing from treatment—and cost—per unit, on to the overall project costs, it is clear that one can quickly arrive at budgets of the order of several million (even several tens of million francs) to be invested for a remediation operation. Thus a chronic fuel leakage in a storage tank resulting in soil pollution over a surface area 30 m × 50 m and 3 m deep, will require an overall budget of 1 to 2 million francs for an 'in-situ' treatment by biodegradation. If the contaminated surface covers a hectare, the cost of operation could rise to 5 to 13 million francs. Dumping an equivalent volume would require around 30 million.

At the stage of estimation of cost of treatment for an entire site, the difficulty lies in precise estimation of the volume to be treated, in terms of quantity as well as concentration and nature of the contaminants present.

Table 6.1: Technologies of remediation—order of magnitude of current costs per unit surface area, volume or mass

Technology	Estimated unit cost
Dumping	500 to 1000 F/t
Pumping-skimming	100 to 500 F/t or m^3
Venting-extraction by Air Vacuum' (for a concentration level of pollutant around 1000 to 100 ppm)	100 to 200 F/t ... over 4 to 8 months
On-site stripping	150 to 300 F/t or m^3
Confinement in safety in specified refuse dumps	4000 to 6000 F/t
Barrier = moulded walls	350 to 1050 F/m^2
= panels	240 to 640 F/m^2
= geomembranes	200 to 500 F/m^2
Roofing in safety with gas collectors	200 to 300 F/t
On-site stabilisation or rendering inert	150 to 1200 F/t
In-situ vitrification	1200 to 2500 F/t
Chemical washing of excavated soils	150 to 600 F/t ... up to 1000 F/t
In-situ washing	200 to 600 F/m^3
Reduction/Oxidation	100 to 600 F/m^3
Prepared as solution/Extraction	500 to 1000 F/t ... up to 8000 F/t (supercritical fluid)
Dehalogenation	sup. at 1000 F/t
Acoustics	Figures not available
Incineration, away from site	1800 to 3000 F/t, up to 7000 F/t for specific substances (PCB etc.)
On-site thermal treatment	400 to 1350 F/t
In-situ bioremediation	100 to 500 F/t
'biopiles'	400 to 1300 F/t
Biodegradation in reactors	300 to 1000 F/t
Composting, landfarming	100 to 400 F/t
Biosparging, bioventing	100 to 300 F/t
Phytoremediation	Figures not available
Humid zones	Figures not available

Broadly speaking, the impact of precision of diagnosis on the final cost of rehabilitation thus becomes a determinant. This is illustrated in Fig. 6.1 in which the cost of treatment of remediation can be seen to decrease significantly with quality and extent of details of the diagnosis, thereby compensating appreciably the extra cost of the larger volume of investigations carried out initially.

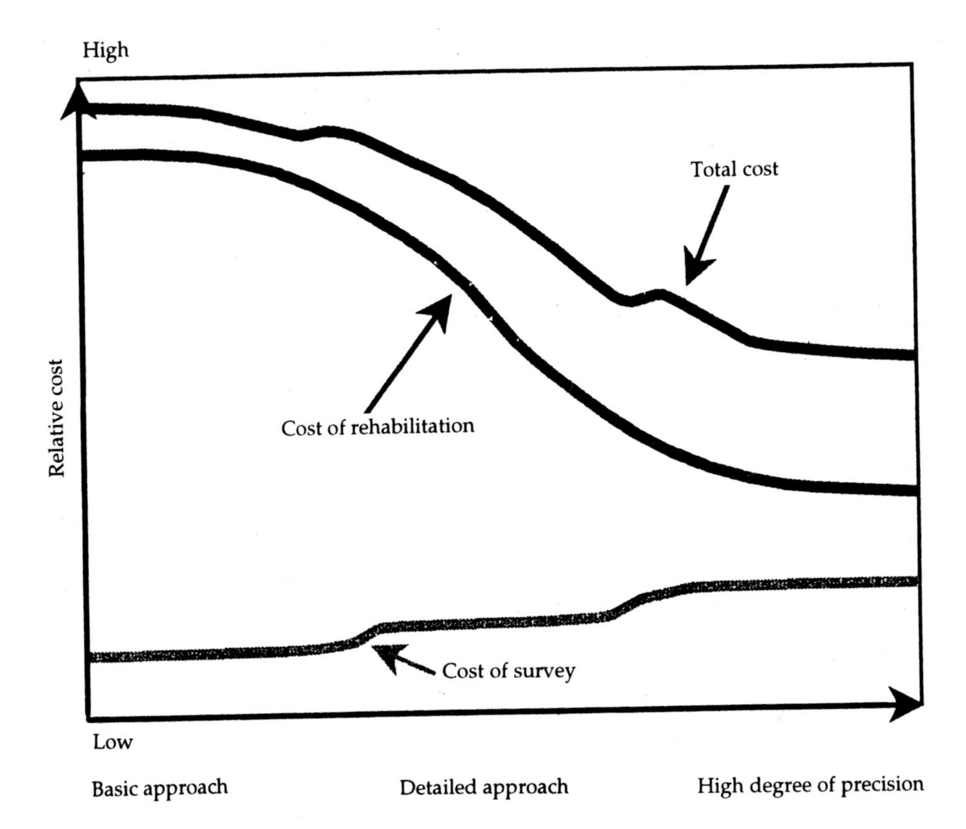

Fig. 6.1: Influence of level of investigation of a site on the cost of rehabilitation.

The following example will give a better understanding of this relation. On an industrial fallow a first diagnosis, relatively quick and rudimentary—termed the 'basic approach' in Fig. 6.1—revealed a contamination of soils extending over 4 hectares and reaching a depth of 5 m. The following substances were reported to be present in 'high' concentration: heavy metals including lead and copper, phenols, mineral oils and tar. The plan of remediation proposed construction of a confining wall around the 4-hectare patch, anchored at 10 m depth in a layer of (impermeable) clay, together with installation of a network of drains upstream for diverting the flow of water away from the contaminated zone. A system for checking growth of the contamination and reliability of confinement, not included in the cost of the operation, would have to be provided and maintained during a period of time, the length of which is a priori indeterminate.

The cost of the operation can be roughly estimated as follows:

— diagnostic phase: 150,000 F;

— construction of moulded wall: 800 m long, 10 m high at the rate of 600 F/m^2, say 4,800,000 F;

— installation of a network of drains upstream of the zone: trenches about 600 m in length with drains for flow of water towards a natural ditch, say 750,000 F;

— **total (excluding the monitoring and control system): 5,700,000 F.**

Later, in view of the high cost of the proposed rehabilitation, a detailed study was undertaken to precisely determine the origin and importance of the pollutions highlighted during the first approach. Its major conclusions led to the following diagnosis:

— lead and copper uniformly contaminate the entire 4 hectares but are present in a stable form (oxide-hydroxide phase) in the temperate humid climate and under conditions of pH-Eh measured in the soil of the fallow;

— oils and tar present around an old workshop contaminate the surface soil over 35 m × 30 m area;

— water table is contaminated by phenols only in top layers and contamination is concentrated at location of old kilns, affecting only 10% of the site.

The plan of remediation proposed:

— Installation of a system of 'in situ' biodegradation by pumping and injecting the groundwater (which would also play the role of hydraulic confinement) at the location of the old kilns to destroy the readily degradable phenols.

— Excavation of about 1750 m^3 (35 m × 50 m × 1 m) of soil contaminated by oil and tar to be treated, depending on their concentration in the polluted earth, either by dumping or conversion in compost piles on the site; the two plans are equally effective.

— Remaining part of the fallow left untreated, the metallic contamination posing no risk for the neighbouring population.

The cost of these new operations can be estimated as follows:

— subsequent phases of diagnosis, 1,000,000 F;

— excavation of 1750 m^3 of polluted earth, composting half of it, say about 330,000 F, 1320 tons at the rate of 250 F/t;

— dumping other half of soiled earth, say about 1,000,000 F, 1320 tons at the rate of 750 F/t;

— in situ biodegradation over a period of 3 to 6 months of 4000 m^2, i.e., of the order of 1,200,000 F for a trench 4000 m^2 in area and 2 m deep at the rate of 150 F/m^3;

— **total: 3,530,000 F**

The difference between the two solutions proposed is large, illustrating the contribution of survey investigations to the performance-price relationship for rehabilitation of a site in toto. Let it also be noted that the second system is definitive and, unlike the first, requires no follow-up action (check-ups on effectiveness) over the course of time.

In determining the orders of magnitude, the following figures have been accepted as representative of costs corresponding to two different levels of investigations and work carried out on a contaminated site:

— documentary survey, several tens of thousand francs;

— diagnosis of terrain, with sampling and analysis: several hundred thousand francs;

— remediation operations: one to several million, sometimes several tens of million francs.

Obviously, this comparison of costs is only in terms of orders of magnitudes and depends entirely on size of the site, its complexity and the risks posed by the contaminating substances present. So, the diagnosis of the terrain of and the site contaminated by a chemical industry covering several hectares would require investigations costing between 1.5 and 2 million francs, while the site polluted by a service station could be investigated for less than a hundred thousand francs.

VII

New Legal Requirements

7.1 FRENCH LAW

In France the legal system has gradually armed itself with a legal arsenal for the protection of the environment. Barring some rare texts of the nineteenth century and the first part of the twentieth, the first legal efforts apropos generation of wastes and installations designated for the protection of the environment appeared only in the 1970s. This was followed by a series of decrees and circulars for compliance concerning points of law. In the last few years the process of finalising codes on regulations has accelerated, especially from the legislative point of view, and environmental protection has taken on a broader, more concrete framework.

The more important legal codes in France, arranged chronologically, are as follows:

— 19 December 1917: law pertaining to dangerous, nuisance-causing and insalubrious establishments;

— 2 August 1961: law on atmospheric pollutions modifying the law of 19 December 1917;

— 15 July 1975: law regarding elimination of wastes;

— 19 July 1976: law pertaining to notified (hazardous) establishments;

— 10 July 1990: bylaw prohibiting discharge into groundwater of substances originating from classified installations;

— 3 January 1992: law concerning water;

— 13 July 1992: law concerning waste management and classified installations, modifying laws of 15 July 1975 and 19 July 1976;

— 1 March 1993: decree concerning classified installations with respect to all types of discharge and abstraction of water;

— 9 June 1994: decree modifying that of 21 September 1977 concerning content of Impact Assessment studies, storage of waste and closure of classified installations;

— Lastly, 16 December 1992: law modifying the penal code, laying down the principle of penal responsibility of enterprises guilty of contaminating the natural milieu. This law, by anticipating an ensemble of measures which might punish just particular individuals, especially, as in the past, the person embodied as head of an enterprise, provides greater strictures in the legal framework for contravention of environmental protection.

This legislative synopsis regroups three main topics of particular interest to us in the framework of protection and remediation of the subsoil and water, namely:

— classified installations,
— wastes,
— management and protection of (surface and ground) water.

7.1.1 Classified Installations

An installation classified for the protection of the environment (ICPE) is one that can pose risks or nuisance for the environment. All industrial or agricultural activities can be included; they may be either in the public or private sector. In France 500,000 installations have been classified as installations needing declaration and another 50,000 'installations needing authorisation', of which about 500 have a 'special hazardous status' termed 'Seveso' due to the particularly dangerous nature of their activities.

The law stipulates that all new activities (new installations, expansions, modifications...) or renovation of a site and in particular a change of occupier, must be declared before the regulatory authority or be authorised by it to be able to operate as a classified installation.

The procedure for applying for authorisation takes a long time (8 to 12 months). The dossier has to be submitted to the regulatory authority by the occupier of a new activity or new site before work can commence. The dossier in particular must include a detailed study of the impact and risks and the plan of treatment of wastes and effluents, giving precise details as to the nature and quantity produced. The dossier will then be technically scrutinised by the competent authority (DRIRE, see Table 7.2) as well as by some departmental, state and municipal bodies. A public investigation (hearing) is carried out before the dossier is submitted to the Departmental Committee for Environment, which will communicate its favourable or adverse opinion, together with technical recommendations for exploitation, to the administrative head. In the case of a favourable opinion, an administrative order will be issued and the necessary permission for construction granted. The order will stipulate from the future exploiter that all the technical specifications (in terms of the precautions to be taken, norms to be observed and management of wastes and effluents) be adhered to.

For a declaration, the procedure is simpler, concluding with submission of the dossier and its technical appraisal by the administration.

The **law of 16 July 1976** stipulates that all new installations must undertake 'impact and risk assessment' studies. The exploiters are obliged to take into account the possible effects of their activities on the site environment. Upon cessation of activities, they must inform the authorities and restore the state of the site.

Compared to the law of 19 July 1976, that of 13 July 1992 is more restrictive in this regard. It enlarges the range of activities in which an authorisation, rather than a declaration, is required and prescribes very strict control and precautions for maintaining ecological balance before exploitation will be allowed. Further, a new activity can only be initiated if the exploiter can show adequate technical and financial capacity to guarantee restoration of the site upon cessation of activities.

The bylaw of 1 March 1993 pertains to 'classified installations' requiring an 'authorisation'. In respect of abstraction and consumption of water as well as various types of discharges, the law is definitive as it fixes the gamut of minimal stipulations that must be honoured with reference to environmental protection, as covered by the laws of 3 January and 13 July 1992 (Table 7.1).

Lastly, **one of the three decrees of 9 June 1994** relating to the definitive closure of a classified installation, stipulates in particular that a site must be restored at the time of definitive cessation of activities, which implies implementation of the measures prescribed by the regulatory authority including elimination of wastes and dangerous substances, remediation of soils and groundwater, a plan of surveillance of the environmental state and so forth. For installations subject to 'authorisation', a memorandum regarding the state of the site and the measures of restoration must be submitted.

To enable enforcement of the various environment-protection regulations, the law of 1992 bestows on the administration (see Table 7.2, the list of administrative offices and public organisations concerned with the environment) broad powers of intervention, especially issuance of complementary decrees, institution of constraints, demanding deposit of monies inclusive of guarantee, and stoppage of activities earlier permitted.

It can equally impose on the persons responsible for the task of rehabilitation of identified pollutions, the decree of 9 June 1994 which reinforces this power. In the case of an orphaned site—one without a known proprietor—the ADEME (Table 7.2) becomes responsible for the operations of remediation.

When an exploiter closes down his activities, he must make a declaration to the regulatory authority and is under obligation to restore the site in

Existing enterprises had to conform to the requirements of the law within three years from the date of promulgation of the law (3 January 1992).

For enforcement on the agencies working in the basins, **two special licenses** and cess/charges have been devised for the occupiers and proprietors of sites:

— 'Abstracting' license: applicable to water used on site. The cess on water used is determined per cubic metre.

— 'Pollution' license: applicable to waste water; the charges are determined by the amount of pollution generated, measured cumulatively by an ensemble of parameters, such as quantities of suspended matter (SS), oxidisable material, organic nitrogen, ammoniacal nitrogen, phosphorus... for the total volume of discharge.

The quantity and quality of water at the inlet and also of the discharge at the outlet are to be measured by the occupier who must render account to the administration, which has the right to verify whatever it considers necessary.

7.1.4 Incentives to Enterprises for Pollution Abatement

While the legislative framework on management of wastes and prevention/resorption of pollution incorporates punitive measures if the rules are not complied with, there are also incentives provided which help the industrialists derive benefits of technical and financial aid when they undertake projects aimed at improving the quality of environment.

Thus, ADEME uses the tax collected on discharges of effluents and wastes to aid industrialists in the installation of devices of treatment or development of innovative low-pollution techniques. Scholarships to persons engaged in scientific research or working on Research and Development projects for an enterprise, constitute part of the ADEME policy for aid.

In a similar spirit, water regulatory agencies may grant financial aid in the form of a loan or credit towards investment as incentive for the protection of ground or surface water resources. Treatment of certain wastes, remediation of soils polluted in the past and erection of new or more efficient installations, constitute some of the projects sponsored by the French agencies in the context of their policy concerning protection of the environment.

Finally, certain regional or general councils are engaged today in providing more and more support to the industrialists in connection with the management of wastes and chalking out plans for treatment.... This aid may be in the form of providing direct or indirect financial assistance: granting loans, reducing charges or relaxing taxes etc.

7.1.5 Conclusion

Although efforts made in the last few years by legislators on the subject of environment have filled to a large extent the legislative void, nonetheless the system still needs to be further strengthened if an efficient and coherent protection of the environment is to be attained. The rules and procedures as existing today are not always adaptable to the situations obtaining in the field and the administration still lacks the means for enforcing compliance.

Also, the French law as of now is not very precise about the division of responsibilities between the State and the industrialists. Who is to pay for what? In what proportion? Similarly, it does not define what pollution is; nor the criteria which would help to characterise contamination with certainty. How to evaluate the likely impact of a pollution and its future ultimate fate? Should the impact assessment procedures be standardised? How to implement the European directives? Should one wait for development of jurisprudence on the subject? In other words, we come full circle to the essential questions posed throughout this book, right from Chapter 2 onwards.

It is clear, however, that the French law on the subject of protection of the environment will become more and more stringent and restrictive under the pressure of public opinion and the media and more concerned with regulations compatible with requirements of the European Community. Right now the legislation focusses on the state of sites and their repercussions on the environment. In future, regulations will be more particularly concerned with the domains of prevention and attaining a 'near zero state of pollution'.

7.2 MAJOR EUROPEAN TRENDS

The European Community proposes and enacts a number of directives covering different topics and domains of protection of the environment.

As on date one can count nearly 250 directives, approved or in the mill, concerned with environment in a broad sense, and covering such aspects as protection of water, transport and treatment of wastes, air and noise pollution, industrial and chemical hazards, conservation of nature... (Environmental Research Newsletter, 1993).

In the European Community procedures a directive constitutes a set of legal recommendations approved by the European Council and is circulated to all the member countries of the Community. Each directive specifies certain objectives and concrete targets to be attained but the member countries are free to choose the means for such attainment and to integrate the goal of the directive into their own national legislation.

7.2.1 Protection of Water

Certain directives, approved or under consideration by the European Council, constitute a definite advance in the protection of soils and water in the framework of the rehabilitation of contaminated sites.

In the domain of water, the following two directives are of particular importance:

— Directive 76/464/EEC of 1976, titled 'hazardous substances', seeks to prohibit or regulate the discharge of certain hazardous substances into surface waters.

— Directive 80/68/EEC, issued in December 1979 and titled 'groundwater', seeks to prevent pollution of groundwater by different categories of harmful substances. It obliges the member states to prohibit direct discharge into the subsoil of a list of chemicals considered highly toxic, and to limit the indirect discharge of a series of other less harmful substances by infiltration. However, the directives are vague and leave a lot of latitude for the member countries in respect of other sources of pollution, e.g. residual chemicals from agriculture etc.

The two directives accompanied by two schedules of the chemicals are available as a booklet.

• The first schedule, called the 'black list', enumerates chemicals considered highly dangerous for the environment. These substances generally carry simultaneously properties of high toxicity, bioaccumulation and long persistence in the medium. Some are known to be carcinogenic, mutagenic or teratogenic. For the first directive, the well-known list of '132 substances' is considered a reference; for the second, the black list is almost identical to that of the first, but a bit more strict, incorporating for example cyanide compounds and hydrocarbons which are non-persistent in the medium. Table 7.3 lists the 9 groups of substances included in these lists.

Table 7.3: Groups of substances included in the 'black list' of the European directives for protection of water

Group no.	Group	Example of compounds
1	Organohalogens	PCB, chloro-BTX, carbon tetrachloride, dichloromethane, aldrine, chlordane, DDT
2	Organophospates	Parathion, malathion
3	Organostannites	Dibutyl salts of tin
4	Carcinogenic substances	PAH, chloro-pyridine benzene
5	Arsenic compounds	As
6	Mercury compounds	Hg
7	Cadmium compounds	Cd
8	Mineral oils and hydrocarbons	BTX, PAH
9	Cyanides	Cyanides

• The second schedule, called the 'grey list', enumerates substances capable of producing harmful effects on the environment, more or less marked, depending on the medium in which they spread. This list contains in particular heavy metals (Pb, Zn, Cr, V...), fluorides, boron, berylium, ammonia, nitrites.... As in the case of the black lists, the differences between the grey lists of the two directives are nominal, the second repeating and completing the items left out in the first.

7.2.2 Abatement and Treatment of Wastes

A number of directives are concerned with the treatment of wastes and reduction of pollution.

Thus directives 74/439, 442/EEC and 91/156/EEC prescribe technical norms for the storage of wastes to be discharged and the obligation of pretreating them before storage so that only the ultimate wastes need be taken into account for storage. French legislation has included these different elements in the new regulations brought into effect in 1992. Similarly, several directives regulate the transport of wastes within and between member states, stipulate the conditions of interfrontier transport of dangerous wastes and prescribe strict procedures of control (directives 84/631/EEC, 86/279/EEC, 87/112/EEC...).

Regulation is not limited to solid wastes. Directive 91/271/EEC of May 1991 concerns treatment of liquid wastes and prescribes the obligation of collection and treatment of effluents before their discharge into the aquatic medium.

Also some directives are directly concerned with reduction of potential pollution. In the domain of agriculture for example, the directive of 12/12/91 (91/676/EEC) restricts use of chemical and organic manures in all cultivated zones designated sensitive, i.e., in which the quantum of nitrates in the subsoil water exceeds 50 mg/l, and stipulates strict compliance with 'good agricultural practices'. Also in the domain of agriculture, directive 86/278/EEC defines the limits of concentration of heavy metals admissible in the manure sludge used as a fertiliser in farms.

Another proposal seeks reduction of the emission of VOC (volatile organic compounds) from the petroleum industry at the stage of storage of products, transport and 'bulk distribution'. The objective is to attain in ten years' time, 90% reduction in the emission of these substances in the atmosphere by employing systems of recovery of vapours. This reduction represents recovery of the order of 500,000 tons of hydrocarbons per year within the European Community.

As for the likely environmental impacts of new activities, directive 85/337/EEC of 27 June 1985 obliges the member states to employ a procedure of evaluation of the impact these activities could cause on the fauna, flora and human beings. It not only stipulates that the natural environment

Another enterprise, following the rupture of a dyke, was fined 600,000 F for damage and interest to the complainant for the pollution caused by the chemical residues of the distillery.

Although at present legal cases pertaining to the environment constitute only two per cent, the judgements are becoming more and more severe and will soon lay stress on need for a new framework of penal procedures (law of 21 December 1992).

Given the fact that French regulations are becoming more and more complex and the rapid evolvement of European legislation, large industrial groups are increasing the measures taken for protection of the environment while the small and medium units as yet mostly lag behind considerably. Anyway, they can no longer ignore the fact that environmental costs today represent 5 to 10% of their business expenses in various taxes.

In view of all these costs, investment on prevention of pollution, protection of environment and training of personnel is still the most profitable solution.

A document published in 1992 by the European Community with reference to the Community's policy and action programme for the environment (Table 8.2) furnishes some examples of the period of return of industrial investments associated with low-pollution technological changes and abatement of environmental degradation.

Table 8.2: Examples of return of investment made on reduction of environmental degradation (extract from the environmental policy and technological progress—source; EEC, 1992)

Industry	Method	% Reduction	Period of amortisation
Fertiliser	Separation of factory wastes	100% dust	10 months
Heavy machines & tools	Chain of procedures	80% sludge and water	2.5 years
Automobile	Procedure of pneumatic cleaning	100% sludge and water	2 years
Electronics	Procedure of vibratory cleaning	100% sludge	3 years
Paints, coatings	Procedure of pneumatic cleaning	100% solvents, paints	less than 1 year
Leather	Ionic separation, adsorption	99% chrome	2 years
Pharmaceuticals	Replacement of organic solvents wth water-based solvents	100% solvent	less than 1 year
Organic chemicals	Adsorption, condensing wastes, airing	95% organics	10 months
Photography	Electrolytic recovery, ion-exchange adsorption	85% developer, 95% fixator; silver, solvents	less than 1 year
Metallurgical and metal-fabrication	Ultrafiltration	100% solvents and oils 98% paints	2 years

IX

Conclusion

During the last few years the topic of 'contamination/remediation of soils and groundwater' has become very sensitive and demands a new area of competence. An entirely new subject, it has attracted attention at the global level from all persons engaged in economic, social or legal professions.

However, it remains as yet a none-too-well-defined domain even though legislation prescribes more and more precisely what needs to be done, implies more and more specific constraints on industrial activities depending on the situation, and formulates administrative measures and necessary procedures. In many cases the decisions to be taken appear vague, sometimes even contrary to the interests at stake.

Nevertheless techniques of rehabilitation do exist. Well-developed and experimentally validated in neighbouring countries, sometimes within the country, they offer curative solutions and many possible modes of treatment, leading to better and better remediation at lower and lower cost. In addition, they have the advantage of rapid evolution, regularly integrating the results of research, literally 'launched' in full tempo. Thus some compounds considered until a few years ago to be resistant to treatment can today be readily degraded; or some methods which earlier required large budgets for their execution, have now become financially competitive.

Similarly, an enhanced awareness of dangers of simple locational transfers of pollution (release of toxic material into the atmosphere, dumping residuals of remediation etc.) has dawned on all parties involved with deeper and marked concern for effectively degrading contaminants by transforming them into harmless substances, rather than simply changing the location of their storage or discharging them in another form.

The panorama of techniques of remediation of polluted sites briefly discussed here is obviously not exhaustive and may over time appear quite obsolete given the rapid development of technologies. Nevertheless it constitutes an inventory of the methods and techniques proposed to date, as well as their possible utilisation and varied degree of experimentation, as an attempt to tackle the situations generated by our past activities.

As a matter of fact, we have been concerned with the tools of remediation and management mistakes of the past—taking care of liability already incurred to be more precise. It is clear that the motto of the future should be prevention rather than 'cure'. The primary objective will thus be to protect and manage our environment in full awareness of the cause, by avoiding repetition of the errors of the past. If our efforts bear fruit, it is certain that 'manuals' of this type will be acquired by archival libraries and housed there permanently.

Photo no. 17 (Source: P. Lecomte)

Suggestions for Further Study and Bibliography

General Reference Works

Académie des sciences—Pollution des nappes d'eau souterraine en France (rapport N° 29, 1991, 150 p.).
ADEME—Sites pollués (Editions de I'ADEME, numéro spécial, mars 1994, 55 p.).
Allègre C.—Economiser la planète (Fayard, 382 p).
Cans R.—La bataille de l'eau (le Monde Editions, 212 p.)
Datar—La réhabilitation des friches industrielles (La documentation française, avril 1991, 56.p).
Destribats J.M., Prez E., Soyez B.—La dépollution des sols en place, techniques et exemples. (Etudes et recherches des laboratoires des ponts et chaussées, série environnement et génie urbain EG10, 1994, 120 p.).
Goldsmith E.—Le défi du XXI^{e}siècle (Editions du rocher, 498 p.).
Goldsmith E., Hildyard N.—Rapport sur la planète Terre (Stock, 380 p.).
Gouverne L.—Histoire d'eau—Enquête sur la France des rivières et des robinets (Calmann-Lévy, 231 p.).
IFEN—L'environnement en France. Edition 94–95 (Dunod, 399 p.).
Ramade P.—Précis d'écotoxicologie (Masson, 300 p.).

Books and Articles Cited in the Text

AFNOR, 1994. Dictionnaire de l'Environnement. Editions AFNOR, 306 p.
Agence Fédérale pour l'Environnement aux USA (US EPA), 1995. Vendor information system for innovative treatment technologies: VISITT Database Version 5, Technology Innovation Office.
Agence Fédérale pour l'Environnement aux USA (US EPA), 1995. Remediation Technologies Screening Matrix and Reference Guide. EPA 542-B-95-005, 142 p.

Anonyme, 1991. Contaminated soil—regulatory issues and treatment technologies. The Hazardous Waste Consultant, Sept.–Oct. 1991, 4.1–4.24.

Anonyme, 1992. Evaluating the feasibility of in situ bioremediation. The Hazardous Waste Consultant, janv./fév. 1992, 1.16–1.20.

Anonyme, 1992. Innovative in situ cleanup processes. The Hazardous Waste Consultant, Sept.–Oct. 1992, 4.1–4.38.

Antoine G., Robert P., Iung O., Ebérentz P., 1993a. Sinistre ferroviaire de La Voulte (07), 13 janvier 1993: Mise en œuvre immédiate des opérations de dépollution. TSM « Pollutec 1993 », septembre 1993, 421–428.

Antoine G., Pellet P., Roche C., Cuisson J., Modolo R., Ebérentz P., Beguin D., 1993b. Sinistre de Chavanay (42): Les grandes étapes de la décontamination. TSM « Pollutec 1993 », septembre 1993, 429–440.

Barrès-Lallemand A., 1993. Guide pratique d'échantillonnage des eaux souterraines. Rapport BRGM R37390, 105 p.

Bonnefoy D., Bourg A., 1985. Fonds géochimiques régionaux et contamination des sols: l'example du bassin versant de l'Orne. Hydrogéologie N° 1, 45–53.

Bouchelaghem A., Magnie M.C., Billard H., 1995. Traitement par stabilisation/solidification de résidus dangereux. Techniques Sciences et Méthodes, N° 5, 407–412.

Bourg A., Mouvet C., Lerner D., 1992. A review of attenuation of trichloroethylene in soils and aquifers. Quarterly Journal of Engineering Geology, 25, 359–370.

Bonin H. Jean P., Philibert P.J., Guillerme M., Fornieles J.M., 1991. Expérimentation de décontamination par biodégradation accélérée in situ au Port aux Pétroles de Strasbourg (Bas-Rhin, France). Hydrogéologie, N° 4, 303–310.

Chevron P., Charbonnier P., Defives C., Lecomte P., 1997. Restoration of a nitrogen-polluted aquifer by biotransformation in situ. 4th Intern Symposium, New Orleans, In: In situ and on-site bioremediation 5: 405–409.

Côme B., Lallemand-Barrès A., Ricour J., Martin S., 1993. Applications comparatives de méthodes d'évaluation de « risques » liés aux sites pollués: premiers enseignements et perspectives. TSM « Pollutec 1993 », septembre 1993, 447–452.

ECETOC, 1990. Hazard assessment of chemical contaminants in soil. Technical report N° 40.

Eickmann T., Kloke A., 1991. Nutzungs— und schutzbezogene Orientierungsdaten für (Schad)stoffe in Böden.—Das Verband Deutscher Landwirtschaftlichen Untersuchungs— und Forchungsanstalten. VDLUFA, heft 1.

Environment Research Newsletter, 1993. Commission of the European Communities, N° 11, SP-I.93.27, 35 p.

EPA, 1990a. Revised Hazard Ranking System—Final rule. Rapport EPA N° PB91-100800.

EPA, 1990b. Handbook on in situ treatment of hazardous waste-contaminated soils. Rapport EPA/540/2-90/002, 157 p.

EPA 1992. Biocontrol soil washing—System for treatment of a wood preserving site. Rapport EPA/540/A5-91/003, 55 p.

EPA, 1993a. Remediation technologies screening matrix and reference guide. Rapport EPA/542/B-93/005.

EPA, 1993b. Superfund Innovative Technology Evaluation Program. Technology Profiles 6th edition. Rapport EPA/540/R-93/526, 424 p.

EPA, 1993c. Field application of bioremediation. In « Bioremediation in the field », N° 9, août 1993, 8–43.

Frost and Sullivan Office, 1993. Réhabilitation des sols contaminés. Rapport E1736.

Géoconfine 1993. Géologie et confinement des déchets toxiques. Actes du symposium international de Montpellier (France), juin 1993, Ed. Arnould M., Barrès M., Côme B. (Balkema Ed., Rotterdam), volume 1, 610 p.

Groo J.P., 1987. Destruction des PCB: Procédés en cours de développement. RGE N°8, 156–162.

Gros D., 1981. Une dépollution qui rapporte. Bulletin d'information de l'-Agence Financière de bassin Rhin-Meuse.

Hushon J.M. et al, 1993. Comparison of hazardous waste site ranking models. The hazardous waste consultant, Sept/Oct. 1993, 1.14–1.17.

Heerling B., Stamm J., Alesi E.J., Brinnel P., 1991. Vacuum vaporizer wells (UVB) for in situ remediation of volatile and strippable contaminants in the unsaturated and saturated zone. In proccedings of Symposium on soil venting, April 1991, Houston (Texas), 26 p.

Jean L., 1997. Le CO_2 supercritique dans le traitement des sols pollués. Environnement & Technique, Info-déchets-Courants, mai, N° 166, pp. 51–56.

Keely J.F., 1989. Performance evaluations of pump-and-treat remediations. Rapport EPA/540/4-89/005, 19 p.

Keith L.H., 1990. Environmental sampling: a summary. Environ. Sci. Technology, 24 N° 5, 610–617.

Lafitte Ph., Maiaux C., Ricour J., Ruquoi D., 1985. Résorption des nuisances engendrées par le « triangle de Carling » (Moselle). Hydrogéologie, N° 2, 133–141.

Lageman R., 1993. Electro-reclamation—State of the art. In the conference documentation of « Contaminated Land », London, Feb. 1993, 12 p.

Lagny C., 1991. Cahier des charges et dimensonnement d'un pilote venting. Note technique BRGM, 91NT016-4S/ENV, 29 p + annexes.

Lapierre J., Mannschott C., Sauter M., 1992. La contamination d'un captage d'eau potable par du tétrachloréthylène. Courants, N° 17, 81–87.

Le Hecho I., Marseille F., 1997. Elaboration d'une banque de données informatisée des technologies de traitement de sols polluées. CNRSSP, Rapport 97/10, 25 p + annexes, non publié.

Mansot J., 1993. Qui risque le plus? Décision Environnement, N° 17, p. 10.